"A WILDERNESS OF DESTRUCTION"

MERCER UNIVERSITY PRESS

Endowed by

TOM WATSON BROWN

and

THE WATSON-BROWN FOUNDATION, INC.

"A WILDERNESS OF DESTRUCTION"

Confederate Guerrillas of East and South Florida, 1861–1865

Zack C. Waters

MERCER UNIVERSITY PRESS
Macon, Georgia

MUP/ H1035

Published by Mercer University Press
1501 Mercer University Drive
Macon, Georgia 31207

27 26 25 24 23 5 4 3 2 1

Books published by Mercer University Press are printed on acid-free paper that meets the requirements of the American National Standard for Information Sciences—Permanence of Paper for Printed Library Materials.

Printed and bound in Canada.

This book is set in Adobe Garamond Pro.

Cover/jacket design by Burt&Burt.

ISBN 978-0-88146-881-6
Cataloging-in-Publication Data is available from the Library of Congress

Contents

"Your country is desolate,
your cities burned with fire;

your fields are being stripped by foreigners
right before you,
laid waste as when overthrown by strangers."

—Isaiah 1:7 NIV

for Maddie

Acknowledgments

The inspiration for this history came more than fifty years ago.

My father worked for the Florida State Board of Health in Jacksonville, but with a growing family, he took a Saturday job as the accountant for a locally owned motel at Waldo, Florida. I often accompanied him on his trips to Alachua County, and one afternoon, as we headed home, my father stopped at one of the state historic markers and told me the story of Capt. J. J. Dickison's cavalry company capturing a Union gunboat on the St. Johns River.

Of course, my childish drawings had the Rebel soldiers swinging on ropes—like pirates boarding a Spanish galleon—but that simple, five-minute tale sparked a fire in me that time has not quenched.

Since that day, I have been determined to tell the story of my home state's Civil War guerrillas, but I began working on this book in earnest only about six years ago. Initial research included four years of reading through hundreds of Northern and Southern newspapers from the 1860s; copying little known, but important, articles by hand; and then typing them for my computer files. As I began writing the rough draft, I spent another two years hunting down obscure and out-of-print books, unpublished theses and dissertations, and various articles with the help of family and friends.

During that time, I was diagnosed with cancer, leading to months of chemotherapy and many more months of physical therapy. In short, I endured a most trying ordeal, but all the while I kept plugging along on the Florida guerrilla project.

My Civil War writing has largely been an attempt to record the role of Florida's Rebel warriors in our nation's War between the States. I was in my mid-teens when my great-grandmother Crumpton—one of Florida's last three widows of a Confederate veteran—passed to her reward, and I grew up on stories of my ancestors' involvement in that conflict. My grandparents, George A. and Mary (Allen) Waters and Zack and Henrietta (Perry) Crumpton, knew the men who fought in the Civil War and heard their accounts of what their parents experienced during that conflict. My father and aunt (Mary Beth Williamson) also shared family stories with a youngster who never tired of hearing those tales.

Often as many as three separate newspaper accounts covered the same battle or action. Pro-Union and pro-Confederate reports usually differed in details, as is to be expected, but both sides often manipulated the facts to shine the best possible light on failures and defeats. With the arrival in Florida of United States Colored Troops, those soldiers and their White officers often added a third (or fourth) perspective to the narratives. I have attempted to quote the various reports, without overt editorial comment, except for one minor incident during the third "annual" occupation of Jacksonville, in mid-1863.

Many people have assisted in my forays into the past. The late Dr. Samuel Proctor encouraged my early writing and accepted my first article for the *Florida Historical Quarterly.* He also gave me a complete set of his series (*Florida a Hundred Years Ago*) that I have often used in my research. I cherish Dr. Proctor's memory as an editor, teacher, and friend.

Dr. David Coles and Don Hillhouse provided invaluable assistance during my early writing and research efforts. Those two historians taught me more about writing than I ever learned in college or in a couple of brief stints as the editor of weekly newspapers. David's PhD dissertation proved an extremely important document and is often quoted in this history.

The *Florida Historical Quarterly* published my first article on Florida irregulars—an essay on John W. Pearson and his "Oklawaha Rangers"—many years ago. During the ensuing years, several of my essays have dealt with the Florida guerrillas in a peripheral way. These included articles on the Confederate commanders at Tampa, the 15th Confederate Cavalry, and a paper on Jacob E. Mickler (primarily written by David Coles) all of which focused, in part, with the state's irregular warriors. More recently, Dr. Seth A. Weitz and Dr. Jonathan Sheppard encouraged me to write a general article on the partisan warfare in Florida for their excellent compilation: *A Forgotten Front: Florida during the Civil War Era* (Tuscaloosa: University of Alabama Press, 2018). That gave me the extra nudge I needed to complete this book.

Numerous friends aided in locating materials used in putting together this history. Nick Wynne, J. Mark Akerman, the late J. J. Koblas, and Robert Redd all gave generously of their time and knowledge. The Cedar Key Florida Historical Society Museum and the Volusia County History Museum (and particularly Robert Redd) were most helpful in providing the information I requested. Dr. Joshua Goodman was amazing in providing information and assistance in obtaining the images for this book. Dr. William F. Stoutamire deserves special thanks for sending a copy of his valuable unpublished thesis, thereby saving me a good deal of time. Kyle VanLandingham, prior to his untimely death, and Cantor Brown Jr., helped make me aware of the importance of the conflict in Florida's cattle country. Of course, Dr. Robert Taylor's pioneering works on Florida cattle and the "Cow Cavalry" emphasized the importance of the southernmost state to the survival of the Confederate armies in Georgia, South Carolina, and Virginia during the last two years of the conflict.

I would be remiss if I did not pay a special tribute to Dr. Joe Knetsch, who read the manuscript and gave a number of helpful suggestions. He also sent me several of his unpublished articles, usually on short notice. I thank him again for all his help with this project.

My brother Robert sent me copies of several difficult-to-locate materials and encouraged me when I felt like burning the manuscript and throwing my laptop in the landfill. I think he will enjoy this book more than most; I suspect the frequent mentions of the fighting in Levy County will remind him of our fishing trips on the Wekiva Run and Waccasassa River.

I thank Lauren Waters and Luke, Amy, and little Maddie Waters for putting up with my strange obsession for telling this story.

My wife, Vonda, has always encouraged my writing. She never complained when family vacations suddenly turned into research trips or when I just had to have "one more book." Without her help, understanding, patience, and love, I would have never had a single book published. She truly blesses my life.

Zack C. Waters
Rome, Georgia

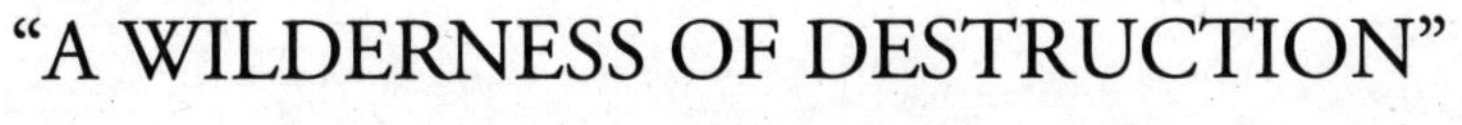

“A WILDERNESS OF DESTRUCTION”

INTRODUCTION

"Our Necessary Resort"

David E. Sutherland, in *A Savage Conflict*, writes, "In proportion to the size of its population, Florida's guerrilla war may have been the most intense in the Confederacy." Despite that appraisal, historians have, almost without exception, treated the state's role in the Civil War as a sideshow rather than a significant part of that conflict. This narrative is an attempt to look at both the magnitude and scope of Florida's partisan warfare.[1]

Florida had been a state barely sixteen years when the Rebels at Charleston, South Carolina, opened fire on Fort Sumter, plunging the nation into civil war. Several weeks earlier, Florida had voted to secede from the Union and throw its lot with the new Southern Confederacy. Though first settled by Europeans in 1565, Florida remained "very much a frontier at the time of the Civil War." In 1860, it had a population of only 140,424; enslaved people composed almost half of the inhabitants of the "Land of Flowers [the state's nineteenth-century nickname]." Florida was ill-prepared to fight a traditional conflict. Though the state was rich in natural resources, there were virtually no manufacturing facilities. As one early historian observed, "Almost everything consumed [by Floridians] except vegetables, forage, and cornmeal was imported."[2]

Much of the state's early settlement had been in the Panhandle and the northern part of the peninsula. Prewar residents typically referred to the state's various regions as West, Middle, and East Florida. "East Florida encompassed the region from the Suwannee River to the Atlantic Ocean. West Florida included everything from the Apalachicola River to the western edge of the panhandle. Middle Florida was made up of the land between the Apalachicola and Suwannee Rivers." A number of large cotton plantations—with slave labor similar to Deep South states such as South Carolina, Georgia, and Alabama—stretched eastward from Tallahassee to Marion County. Most early settlers, however, were "yeoman farmers and cattlemen [who] had moved into the peninsula, lured by the free land offered by the Armed

---

[1] Sutherland, *Savage Conflict*, 257.

[2] Coles, *Florida Civil War Heritage Trail*, 2; W. W. Davis, *Civil War and Reconstruction in Florida*, 33.

Occupation Act.... [They often] led a hardscrabble existence of subsistence farming and cattle ranching."[3]

South Florida—a region including Hernando, Hillsborough, Manatee, Polk, Brevard, and Dade counties—would play an increasingly important role during the last two years of the war. Historian David Coles described that frontier district as "arguably the most remote area in the eastern United States," and a sizeable number of that area's residents apparently hoped that the "protecting barrier of swamps and blue water" would shield them from the coming conflict. Fewer than 8,000 people lived in South Florida, most of them in and around Tampa and Union-occupied Key West. Just 1,183 hardy individuals resided in the countryside and in the "few small towns concentrated around current or former military outposts. Some sugar production took place in the Manatee River area, and the southern region was also the home to thousands of cattle that ranged the scrublands and swamps." As the War Between the States progressed and intensified, those half-wild longhorns would bring the scourge of warfare, death, and destruction to South Florida.[4]

A minority of the state's White settlers failed to share the Confederacy's values, attitudes, or cultural mores, but Florida's response to the new nation's call to arms indicated a general willingness to fight. It was the boast of prior generations of Florida "Crackers" as descendants of the state's White Southerners were called, that their state "contributed more than 15,000 [men] to the Confederate War effort. While this was a small number when compared with other Southern states, it was the highest percentage of available men of military age from any Confederate states." Most of those volunteers enlisted in infantry companies. The 1st, 3rd, 4th, 6th, 7th, and 1st Florida Cavalry (dismounted) regiments fought with the Army of Tennessee; the 2nd, 5th, 8th, 9th, 10th, and 11th saw action in the Army of Northern Virginia. Not all troops from Florida, however, served as foot soldiers. The Marion Light Artillery campaigned outside the state with the Confederacy's Western Army; and a handful of horse companies from West Florida rode with the 15th

---

[3] Coles, *Florida Civil War Heritage Trail*, 2–3; Hawley, "Florida's Civil War Soldiers," 2–4; Wynne and Taylor, *Florida in the Civil War*, 7–12; see generally, Knetsch and George, "Problematical Law," 63–80.

[4] Coles, "Cattle Wars," 95–96; Covington, *Story of Southwestern Florida*, 140. In 1861, Hernando County included three modern counties, Citrus, Hernando, and Pasco.

Confederate Cavalry, operating primarily in Alabama, Mississippi, and West Florida.[5]

The first months of 1862 brought the new nation little but news of defeat, despair, and dashed hopes. The Rebels' campaign to seize the gold- and silver-rich territories of New Mexico and Arizona ended in failure at the Battle of Glorietta Pass and an agonizing retreat back to Texas. The Yankees also won a hard-fought struggle at Elkhorn Tavern (Pea Ridge), thereby temporarily taking control of Missouri and much of Northern Arkansas. A series of even worse debacles occurred in the Volunteer State. By the end of February 1862, Union troops had captured Forts Henry and Donelson, occupied Tennessee's capital city, and forced Gen. Albert Sidney Johnston's graycoats to retreat into Northern Mississippi. On the Atlantic shoreline, bluecoats threatened the islands and outposts along North Carolina coast; Maj. Gen. George B. McClellan, with an army of more than 100,000 men, began a halting advance up the Yorktown Peninsula toward Richmond. It seemed to many people, in both the North and South, that the Federals were on the verge of ending the rebellion before it had fairly begun.[6]

In response to those setbacks, on 1 March 1862, Gen. Robert E. Lee informed Brig. Gen. J. H. Trapier, commanding Southern forces in Florida, of the Confederate government's decision to withdraw virtually all Rebel troops from Florida. Lee, serving as Pres. Davis's chief military adviser, had twice visited Amelia Island and concluded that Florida's "17,000 miles of coastline and its 54,000 square miles of sparsely settled forests and swamps were not considered strategically valuable, nor, for that matter, even defendable." He added, "The recent disasters to our arms in Tennessee [and elsewhere] forces the Government to withdraw forces in defense of the [Florida] seaboard. The only troops to be retained in Florida are such as may be necessary to defend the Apalachicola River, by which the enemy's gunboats may penetrate far into the State of Georgia." The logic of Lee's military decision could hardly be disputed, but to Floridians, it indicated their lack of importance to the new nation. Florida Gov. John Milton understood the implications of this pronouncement better than most. He bitterly complained to Sec. of War Judah P. Benjamin that "the effect of the order is to abandon

---

[5] Hawk, *Florida's Army*, 95–96; Florida Department of State, "Florida in the Civil War"; Waters, "Enigmatic Colonel Harry Maury," 8–17.

[6] Hannings, *Every Day in the Civil War*, 113–69.

Middle, East, and South Florida to the mercy or abuse of the Lincoln Government."[7]

The implementation of Lee's directive affected the state in several ways. Modern historians generally assert that abandonment by the Richmond government weakened the state's devotion to the Confederacy and encouraged many residents to embrace Union policies. A sizable number of citizens, particularly in East Florida, did swear fealty to the Union, essentially sitting out the war. Early Federal reports from occupied Fernandina and St. Augustine indicated that things along the coast "are progressing finely." Union officers announced the [Southern] men "were all renewing their allegiance, but the women were as kantankerous [cantankerous] rebels as ever, or at least fully two-thirds of them." In many cases, the men's "loyalty" seems to have been more a subterfuge to survive occupation than whole-hearted enthusiasm for the United States. Their lack of devotion to the Union dawned slowly on the Yankee officers, but by 1864, a Federal general at Jacksonville bitterly complained that "the majority of refugees [White citizens within Northern lines] in Florida have apparently done us more harm than good."[8]

Runaway slaves—called "contrabands" by the bluecoats—provided far greater assistance to the Yankee war effort than all Florida's lukewarm Unionists combined. Blacks fleeing to the Federals represented a kind of "double whammy" to the state's Rebel sympathizers. The fugitive slaves deprived the Southern planters of their primary source of labor, and many of them joined Federal military units to fight for their freedom. Large numbers of Florida's African Americans enlisted in the United States Colored Troops (USCT). Other Black men served in the Union Navy. Though they rarely appeared in the officers' reports, the "superintendent of the Naval War Records reported that African Americans constituted one-fourth of the [Federal Navy's] Civil War enlistments." It has been estimated that one thousand Florida contrabands joined the North's military forces though those guesses appear to be too conservative. Even when the runaway slaves did not enlist, they helped

---

[7] U. S. War Department, *The War of the Rebellion: A Compilation of the Official Records of the Union and Confederate Armies* [henceforth *ORA*], ser. 1, vol. 6, 402–403; Reiger, "Florida after Secession," 128–31.

[8] Reiger, "Florida after Secession," 133–36; "Further from the Coast," *Charleston [SC] Courier*, 9 April 1862, 1; Schafer, *Thunder on the River*, 232–33.

the Federal cause by serving as guides or providing valuable intelligence to Union soldiers and sailors.[9]

The Davis government's removal of most Rebel troops from the state forced Florida's Confederates to adopt the age-old strategy of the underdog—guerrilla warfare. The culture of the Deep South made partisan operations seem an appealing alternative to a great number of Rebel sympathizers. A recent study of the "shadow war" noted, "Long before the Civil War, guerrillas and guerrilla warfare had been popularized and romanticized in Southern Literature." Writers portrayed "the South [as a region] 'defending its unique way of life through the chivalric code like a medieval knight, a high-bred planter trained in arms and horsemanship [who] would use partisan tactics to repel invasion, inspire, and unify the Southern people.' Such writings helped create the myth of guerrilla warfare as a "patriotic adventure, and above all, glorious."[10]

In addition to the literary propaganda, many Floridians knew from personal experience how the Seminole Indians, using hit-and-run tactics, stymied the United States Army during the long years of the Second Seminole War. Even the state's geography reflected the glorification of the guerilla. The citizens of Marion County, established in 1845, adopted its designation to honor Revolutionary War hero Francis Marion, South Carolina's legendary "Swamp Fox." Adjacent Sumter County, created in 1853, took its name from another Southern partisan—Thomas Sumter, the "Carolina Gamecock." A Floridian writing to a Macon, Georgia, newspaper in late 1861 summed up the opinion of many of the state's citizens. "In the defence of the Florida coast," he wrote, "great assistance can be derived from guerrilla companies [which] would be invaluable.... If our words [woods], which are very thick and heavy, became infected with guerrilla companies, the Yankees, if they attempt to march through the country, will conclude that Florida is a wilderness of destruction."[11]

Despite this widespread belief in the efficacy of irregular warfare, the administration of Pres. Jefferson Davis initially opposed the use of partisan

[9] Buker, *Blockaders, Refugees, & Contrabands*, 42–46; Coles, *Florida Civil War Heritage Trail*, 18.

[10] Ramage, *Rebel Raiders*, 54–66; Sutherland, *Savage Conflict*, 26–32.

[11] Sutherland, *Savage Conflict*, 26–32; Florida Department of State, "County Name Origins"; J. D. P., "King Cotton," *Macon [GA] Telegraph*, 3 October 1861, 2.

combatants. In mid-1861, Sec. of War Judah P. Benjamin unequivocally declared: "Guerrilla companies are not recognized as part of the military organization of the Confederate States, and cannot be authorized by this department." This foolish policy almost certainly reflected the personal beliefs of the chief executive. As a graduate of West Point, Davis fervently believed that regular troops provided the new nation's best chance of battlefield success. A historian of the bitter bushwhacker war in Missouri sarcastically noted, "Jefferson Davis did not believe in guerrilla warfare. Such service was too disorganized, too free of restriction and control, too lacking in petty detail for him to recognize its merit." However, as more territory fell to the enemy, Davis reluctantly modified his stance on the use of the brush fighters. In mid-1863 he wrote, "Irregular bodies of troops are not as efficient as disciplined and instructed soldiers, but when we have not enough of the latter, the former is our necessary resort." This decidedly lukewarm endorsement aside, the Davis never seemed to recognize the value of those who fought the "shadow war."[12]

By contrast, Gov. John Milton became an early proponent of partisan warfare. In 1862, responding to the latest demand from Richmond for yet another Florida regiment, Milton wrote (with undisguised contempt), "Nevertheless enough companies responded to make a Regiment of infantry...[meeting] the requisition of the War Department. But I repeat what I have before said, one thousand men, divided into small companies—well-armed—and acting as Guerrillas or Rangers—and ably commanded—can do more to defend Florida from the enemy than thousands in regular service."[13]

Despite the romantic drivel of nineteenth-century novelists, Florida's partisan combat had few men gallantly dashing about in plumed hats on blooded steeds. A guerrilla campaign has always represented a desperate gamble. The word "guerrilla" comes from the Spanish for "little war," in honor of the Spanish irregulars who harassed the French Legions on the Iberian Peninsula during the Napoleonic War. Despite this relatively modern name, "guerrilla warfare was the earliest form of warfare," predating recorded history. As historian Max Boot notes, "No belligerent in their right mind would voluntarily undertake it [irregular warfare] if there were any credible alternative. Guerrilla...tactics...always have been the resort of the weak against the

[12] *ORA*, ser. 4, vol. 1, 1008; Sutherland, *Savage Conflict*, 53–54; Brownlee, *Gray Ghosts of the Confederacy*, 77.

[13] *ORA*, ser. 1, vol. 52, pt. 2, 489; King, "Shadow Warriors of the Confederacy," 28–30; John Milton, 23 April 1862, Milton Letterbook.

strong. That is why insurgents wage war from the shadows; if they fought in the open, like a regular army, they would [soon] be annihilated."[14]

Irregular warfare invariably frustrated regular combatants. In 1863, a Union officer attempted to delineate for Federal soldiers the role and treatment of Rebel guerrillas. "The partisan leader commands a corps whose object is to injure the enemy by action separate from that of his own main army," he wrote. He continued,

> the partisan acts chiefly upon the enemy's lines of connections and communication and outside of or beyond the lines of operations of his own army, in the rear or on the flanks of the enemy. Rapid and varying movements are the chief means of his success; but he is part and parcel of the army, and, as such, considered entitled to the privileges of the law of war, so long as he does not transgress it.[15]

That statement contained an enlightened attitude toward enemy guerrillas, but the common Yankee soldier found it difficult to relate those words to the foe they fought. A Union combatant campaigning in Northern Alabama spoke for virtually all of his comrades in blue when he declared,

> [T]he rebel cavalry companies...call themselves partisan rangers, but [also] claim to be part of the Rebel army, and expect all of the immunities of civilized warfare. They wear citizens clothes, seize all the property that falls their way, and sharpshoot our pickets when they [can] succeed on slipping up on them...[and then] detach themselves, to prowl, rob, and murder as before. The only treatment for those cutthroats is to shoot them on sight and hang them when taken prisoner.[16]

The tactics employed by irregular warriors almost invariably involve surprise, hit-and-run attacks, sabotage, raids, and greater mobility than traditional military units. Their success also required receiving support from the local populace, superior knowledge of the terrain, avoiding confrontation with larger forces, and fighting smaller, detached groups. Florida's partisan companies proved a particularly popular and effective option to Rebel sympathizers there. A recent study concluded, "The cavalry raider was an

---

[14] Boot, *Invisible Armies*, xxii–xxiv; King, "Shadow Warriors of the Confederacy," 195.

[15] *ORA*, ser. 1, vol. 22, pt. 2, 238.

[16] "Letter from Corinth," *Philadelphia Inquirer*, 6 November 1862, 2.

important part of Confederate military history, as they could move fast, strike hard, then fade away in the face of pursuit.... Their missions tied down thousands of Union troops, disrupted [enemy] communications and transportation, and destroyed millions of dollars worth military property."[17]

[17] Kozikowski, "Guerrilla Warfare: Hometown Heroes and Villains"; Wynne and Taylor, *Florida in the Civil War*, 85–86, 134; Brownlee, *Gray Ghosts of the Confederacy*, 28.

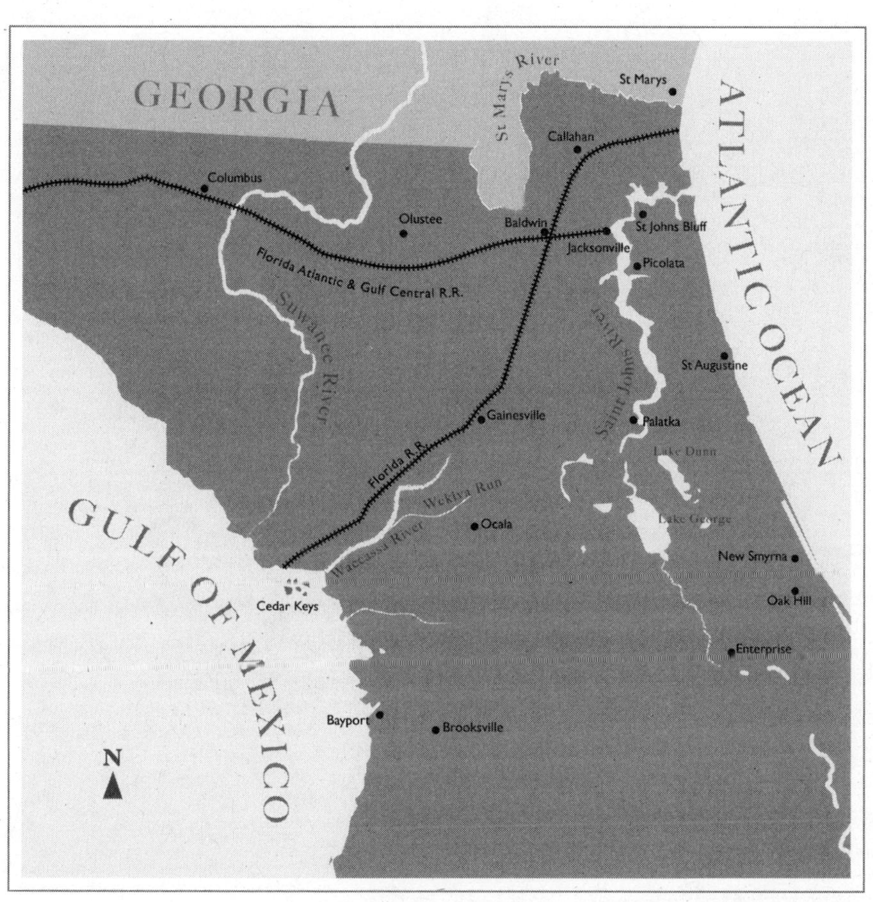

Northeast Florida

Map by Dave Helton

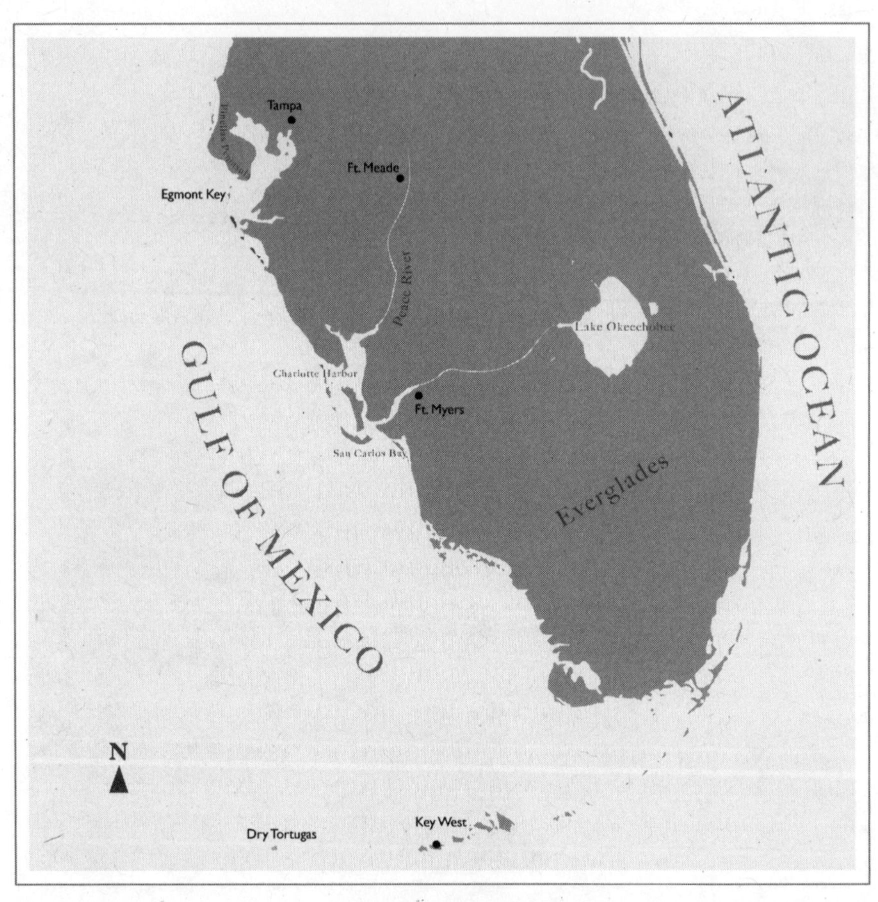

South Florida

Map by Dave Helton

Governor John Milton, a dedicated Rebel,
governed Florida during the war years.

Courtesy State Archives of Florida

Capture of the Gunboat Columbine, sketch from Dickison and His Men.

Courtesy State Archives of Florida

After Union troops took control of the Mississippi River, Florida became the primary source of beef for the Confederate armies.

Courtesy State Archives of Florida

Fernandina fell to the Federals in early 1862.

Courtesy State Archives of Florida

Fort Marion, at St. Augustine, remained in Union hands throughout most of the war.

Courtesy State Archives of Florida

Guerilla leader, Capt. J.J. Dickison became a hero to Florida's pro-Confederate citizenry.

Courtesy State Archives of Florida

Battle of Gainesville, sketch from Dickison and His Men.

Courtesy State Archives of Florida

Battle of Bridge No. 4, sketch from Dickison and His Men.

Courtesy State Archives of Florida

United States Colored Troops served with distinction
during the last two years of the war.

Courtesy State Archives of Florida

Postwar image of Levy County guerilla leader,
Capt. E. J. Lutterloh.

Courtesy State Archives of Florida

Chapter 1

"Bands of Pirates and Robbers"

(1861)

CHATTAHOOCHEE ARSENAL (MIDDLE FLORIDA)

Florida's guerrilla war began with a bizarre surprise attack along the banks of the Apalachicola River.

On 3 January 1861, the state's Secession Convention convened at Tallahassee to determine if the southernmost state would sever its ties with the United States. Madison Starke Perry, Florida's fourth governor, had orchestrated the entire process. A South Carolina native, an early disciple of the Palmetto State's Sen. John C. Calhoun, and an Alachua County cotton planter and slave owner, Perry had stacked the deck to achieve his goal. He made certain that most of the convention's representatives "were members of the planter class, slave holders who operated large plantations given to the production of cash crops, such as cotton and sugar.... Slavery and the protection of slavery were paramount to them, and few were willing to tolerate any individual or group that threatened that institution." As an added precaution, Perry structured the rules of the assembly so that its decision would be binding, without any requirement for a statewide vote to approve the resolution. After a week of debate, on 10 January 1861 the secession ordinance passed by a vote of 62 to 7. By then, however, the state had already commenced military operations.[1]

Nine hundred miles to the north, in Washington, D.C., Florida senators David Levy Yulee and Stephen Mallory, both lukewarm secessionists, had been searching for ways to help their constituents during this time of upheaval. The two lawmakers had developed a network of "Southern men" in the halls of Congress and the US War Department who kept the Floridians informed regarding Pres. James Buchanan's plans. On 4 January, Yulee submitted a request for an inventory of military supplies stored in the Federal

---

[1] Coles, *Florida Civil War Heritage Trail*, 2–3; Wynne and Taylor, *Florida in the Civil War*, 7–12; Proctor, *Florida: One Hundred Years Ago*, 10–13 January 1861 (henceforth, Proctor, *FL100*, and date); Knetsch, "Madison Starke Perry and the Spirit of Secession."

Arsenal at Chattahoochee, Florida. Suspecting the reason behind Yulee's application, US Sec. of War Joseph Holt, a Kentuckian and ardent opponent of slavery, refused to divulge that information. Instead, Holt sent a telegraph to the caretaker at the arsenal to destroy all munitions stored at Chattahoochee—especially the gunpowder. Fortunately for the Floridians, an unidentified telegrapher at Atlanta, Georgia, intercepted Holt's message. The Georgian apparently pigeonholed the communiqué to Chattahoochee, and, instead, forwarded the information to Gov. Perry.[2]

Perry took immediate steps to capture the facility and its valuable stores. The arsenal, completed in 1839, sat on several acres of high ground along the Apalachicola River. The installation consisted of seventeen buildings surrounded by a brick wall, twelve feet high and thirty inches thick. Properly manned, the stronghold should have been able to withstand all but the most determined assault by a large body of well-armed and highly motivated soldiers. Unfortunately for Sec. Holt and his Unionist allies, the entire garrison consisted of a sergeant and three privates.[3]

According to a period newspaper, "Col. W. J. Gunn, of the Seventh Regiment of Florida Militia, received orders from Gov. Perry to proceed at once to the United States Arsenal at Chattahoochee.... The Young Guard [a Gadsden County volunteer company] was ordered out on the errand." Gunn, a wealthy merchant and state militia officer, must have had complete confidence in Perry because "the occupation of U. S. forts and arsenals by states still in the union constituted treason."[4]

At six a.m. on 6 January 6, Gunn's men arrived at their destination in buggies and buckboards. Sgt. Edwin Powell, the arsenal's senior caretaker,

---

[2] Cox, "Apalachicola Arsenal"; "The United States Arsenal at Chattahoochee Taken by State Troops," *Charleston [SC] Mercury*, 15 January 1861, 2. Sen. Yulee would spend seven months in a federal prison postwar for his part in this action (though the real reason appears to have been President Andrew Johnson's extreme anti-Semitism). See Harrison, "The Two Presidents Johnson and the Jews."

[3] Cox, "Apalachicola Arsenal"; Coles, *Florida Civil War Heritage Trail*, 26.

[4] "United States Arsenal at Chattahoochee Taken by State Troops," *Charleston [SC] Mercury*, 15 January 1861, 2; Hartman and Coles, *Biographical Roster of Florida's Confederate and Union Soldiers* (hereafter cited as *BRF*), 1:50; Coles, *Florida Civil War Heritage Trail*, 26; Cooper, *We Have the War Upon Us*, 130. There is considerable debate regarding the leader and unit who took the arsenal. Dr. Coles and others identify the unit as the "Quincy Guards" commanded by "Col. Duryea." See also Proctor, *FL100*, 5 January 1861.

met the intruders at the facility's entrance and launched into a bitter tirade against the "traitors." He reportedly told Gunn that if he had just a few more men, "they would [have] never entered the gates of the Arsenal without first marching over his dead body." He then offered his sword to a Capt. Jones, who supposedly returned the weapon with the gracious comment, "You are too brave a man to disarm." The facility contained "over 5,000 pounds of gunpowder, over 173,000 small arms cartridges, 57 flintlock muskets, and one six-pounder cannon with over 300 shot and canister [shells]." Col. Gunn quickly dispatched the captured munitions to Tallahassee and dispersed his company with a stern warning to never tell a soul what they had done.[5]

St. Augustine

The oldest city in the United States, St. Augustine had a population of slightly more than 1,900 people according to the 1860 Census. Known as the "Ancient City," it also boasted a busy harbor that not only brought in goods and merchandise but also had welcomed a growing number of Northern tourists during the winter. A newspaper, discussing St. Augustine's commercial attributes, reported, "The last place of importance on the Atlantic coast of the southern states is St. Augustine. The entrance to the harbor is in two inlets with only five or six feet of water at the bar at low tide. The [primary] harbor is commanded by Fort Marion, an old Spanish work, which has been recently renovated."[6]

Before the secession crisis, the US Army had stored a large number of cannon at the St. Augustine fort, and Gov. Perry wanted those artillery pieces. Immediately after receiving news of the successful raid on the Chattahoochee Arsenal, Governor Perry directed George Couper Gibbs "to proceed to St. Augustine, to seize the forts, arsenal, and ect. [*sic*]" Gibbs, a dedicated secessionist, commanded a volunteer artillery company at Fernandina. He had also served during the Mexican War as an officer in a Louisiana regiment, and boasted a distinguished lineage that included famed Anglo-American artist J. A. M. Whistler. Perry instructed Gibbs to "act with secrecy and discretion" but authorized him to use whatever force he deemed necessary to capture and

[5] Coles, *Florida Civil War Heritage Trail,* 26; "The United States Arsenal at Chattahoochee Taken by State Troops," *Charleston [SC] Mercury,* 15 January 1861, 2; "Too Brave a Man to Disarm," *Kalamazoo Gazette,* 21 December 1893, 7.

[6] "St. Augustine, Florida: Demographics"; "Southern Harbors of the United States," *Amherst [NH] Farmer's Cabinet,* 20 September 1861, 1.

hold Fort Marion. Taking his own company and a number of men from Capt. John Lott Phillips's "St. Augustine Blues," Gibbs's task force approached the fort and directed Ordinance Sgt. Henry Douglas to surrender the bastion. Douglas, the only Federal soldier at the Ancient City, immediately bowed to the reality of his situation but requested a written inventory of the items "seized." In an article to a postwar newspaper, a participant described the action in a single sentence: "We took possession of the fort," he wrote, "and captured the entire garrison, consisting of one lonely sergeant, well advanced in years, who surrendered very graciously."[7]

Gibbs's militiamen initially worried about the tremendous amount of work that would be involved in emplacing all the cannons at St. Augustine. However, Gen. Robert E. Lee, commanding Confederate coastal defenses in East Florida, Georgia, and South Carolina, had other plans. He ordered Gibbs to dispatch four guns to Fort Clinch on Amelia Island. Four additional guns were emplaced at the mouth of the St. Johns River to guard East Florida's major inland waterway, and the rest were scattered to important points throughout Lee's district. A relieved Rebel soldier at St. Augustine wrote, "Fort Marion was left with only five cannon [which were] mounted in the water battery to defend the city."[8]

An event on 18 August both shocked and saddened St. Augustine's Confederate sympathizers. The Rebel privateer *Jefferson Davis*, which claimed the Ancient City as its home port, floundered and sank while passing the harbor's bar. In her short life as a "sea wolf"—just less than three months—the *Jefferson Davis* had taken the war to the New England coast. Like a seagoing guerrilla, the sleek black privateer's hit and run tactics threw the Northern port cities into a panic unseen since the War of 1812. In addition to causing "the greatest consternation and excitement in Northern ports," the *Davis* captured six prizes totaling almost a quarter of a million dollars. A recent historian reported, "The success of the *Jefferson Davis* represented the height of [Confederate] privateering activity and her total loss...signaled the decline of Rebel

---

[7] "Our Florida Correspondence," *Charleston [SC] Mercury*, 28 January 1861, 1; Allardice, *More Generals in Gray*, 100–102; Coles, "Ancient City Defenders," 73–74; J. Gardner, "St. Augustine in Wartimes," *St. Augustine Evening Record*, 15 July 1914.

[8] East, "St. Augustine during the Civil War," 75–76.

maritime freebooting." Despite the vessel's misfortune, all hands survived the ordeal, and St. Augustine residents treated the crew as conquering heroes.[9]

The actions of Paul Arnau played an important but largely forgotten role during the first year of the War between the States. In 1861, the St. Augustine native and inveterate Rebel, held several important posts, including head of the US Custom House, superintendent of lighthouses, and mayor of St. Augustine. Arnau recognized that "Union military forces needed [the] lighthouses...to facilitate naval operations and delivery of troops and supplies to front-line units near the coast." The signal beacons were especially important because many of the Northern mariners had little knowledge of the Florida's coastlines. A Yankee officer, early in the conflict, averred that the loss of the Southern lighthouses, for a single night, "would be disastrous."[10]

Arnau decided he could best serve the Confederacy by darkening all the lighthouses along Florida's Atlantic coast. He began his campaign in his hometown, aided by Maria Mestre de los Delores Andreu, the keeper of the St. Augustine light. Arnau had appointed Mrs. Andreu to that position after her husband fell to his death while painting the signal tower. (The St. Augustine Maritime Museum notes that Mrs. Andreu "was the first American woman to be made the keeper of a light station in Florida, and the first Hispanic American to command a Federal shore installation.") On the evening of 22 August 1861, the two dedicated Confederates dismantled the St. Augustine beacon and hid the lighting apparatus in an orange grove along the Banana River.[11]

Arnau and Maria Andreu quickly organized a group of like-minded Rebels, nicknamed "the Coast Guard," to darken the other two signal towers along Florida's east coast. The "Coast Guard" easily disabled the Jupiter Inlet lighthouse (in modern Palm Beach County) but faced a sterner challenge at the Cape Florida light (in modern Dade County). The two keepers were dedicated Unionists, but Arnau's small force tricked them into opening the

---

[9] "The Jefferson Davis," *Charleston [SC] Courier*, 2 December 1861, 4; "Jefferson Davis, Privateer," *New Orleans Times-Picayune*, 20 December 1903, 13; Browning Jr., *Success Is All That Was Expected*, 20. See generally, Redd, *St. Augustine in the Civil War*, 49–58.

[10] Thiesen, "The Long Blue Line: Lighthouse Service during the Civil War"; East, "St. Augustine during the Civil War," 76–77.

[11] East, "St. Augustine during the Civil War," 76–77; information on Mrs. Andreu provided in the "St. Augustine Lighthouse & Maritime Museum Visitor Map."

stronghold's tower, and the final lighthouse on Florida's east coast soon went dark.[12]

By 1 September, Union blockading vessels had begun intermittent patrolling of the sea outside St. Augustine's primary harbor entrance. Early accounts indicate that the Federals' initial attempts to halt the blockade-runners proved largely ineffectual, with the shore battery and guns at Fort Marion helping to keep the enemy ships at a safe distance. An unidentified Northeast Florida planter provided a description of a "typical encounter" to a New Orleans newspaper. He reported, "[A] vessel laden with coffee and fruit...[had recently] run the blockade...under heavy fire from shot and shell from the [US] fleet. They [Union artillery shells] all fell short of her, however, and did no damage.... The Floridians express themselves as very confident that, with the assistance of their batteries, they can bid defiance to all and any of Lincoln's piratical invaders."[13]

NEW SMYRNA AND VOLUSIA COUNTY

The hamlet of New Smyrna, founded in 1768 by Dr. Andrew Turnbull, played a much larger role in the Civil War than its small size and isolated locale might have suggested. Turnbull, a physician, landowner, and friend of the governor of British East Florida, had brought more than thirteen hundred settlers to his estate, including many from the Mediterranean island of Minorca. The virgin soil at New Smyrna initially yielded several bumper crops of indigo and sugar cane. Turnbull and his overseers, however, treated the workers so badly that virtually all of them left *en masse*, relocating to St. Augustine. Disease and attacks by Native Americans further depleted the number of settlers who remained at New Smyrna. By 1861, the village consisted of only the spacious Sheldon Hotel and a few houses scattered through the nearby Volusia County woodlands.[14]

---

[12] East, "St. Augustine during the Civil War," 76–77; Dillon Jr., "Gang of Pirates," 441–57; Du Bois, "Two Lighthouse Keepers," 41–50.

[13] "The Blockade of the East Coast of Florida," *New Orleans Times-Picayune*, 11 September 1861, 2. For a differing account of conditions at St. Augustine in 1861, see Redd, *St. Augustine in the Civil War*, 23.

[14] UHP staff, "Caught in the Middle: New Smyrna"; Sweett, *New Smyrna*, no page numbers (n.p.). Regarding the Minorcans, see generally, Griffin, *Mullet on the Beach*.

Prior to the war, Volusia County's primary nonfarming industry involved harvesting hardwoods to supply the US Navy with planks to build and repair ships. For three generations, the Swift Brothers, headquartered in New Bedford, Massachusetts, had employed a large number of locals to cut and season live oak and cedar lumber. However, when Florida voted to secede from the Union, the Swifts departed to New England, leaving behind "huge piles of timber…at [nearby] Live Oak Point."[15]

In mid-May 1861, Confederate officers at St. Augustine received intelligence that several US ships had anchored near New Smyrna to "steal" the hardwood planks. Lt. John C. Buffington, a Georgia native who had been living at Fernandina before the conflict began, volunteered to take a thirty-man squad to Volusia County to capture, or drive off, the timber thieves. Buffington divided his small force, leading fifteen horsemen overland to Live Oak Point while the rest of his company proceeded to Volusia County by sea. By the time Buffington's Rebels arrived, most of the enemy vessels had departed fully loaded with their valuable cargo. A newspaper reported the Confederate "volunteers succeeded in capturing the [Federal] steam tow, the *George M. Bird*, and [also in] securing about 20,000 [feet] of live oak [lumber]." Despite their lack of success in capturing the Federals, when his weary troops returned to St. Augustine, the guns at Fort Marion boomed a salute to Buffington and his men for their "successful mission."[16]

POLITICIANS AND OFFICERS

In October 1861, after almost a year as Florida's "lame duck chief executive," Madison Starke Perry accepted the colonelcy of the 7th Florida Infantry and led his troops to East Tennessee. His successor, John Milton, "a Jackson County planter, lawyer, and politician," soon took office in Tallahassee. Milton loved his adopted state and passionately embraced the ideals of the Confederacy, but his ego kept the state in turmoil for several months after he moved into the governor's mansion. Most of Florida's regimental officers had been what Milton derogatorily referred to as "Perry's Pets," and he seemed determined to purge those men from leadership roles. As a modern historian

[15] Sweett, *New Smyrna*, n.p.; Hebel, *Centennial History of Volusia County*, 3–4; Bowden, "Volusia's Bloodiest Civil War Battles."

[16] Hartman and Coles, *BRF*, 3:1030; "A Seizure in Florida," *Charleston [SC] Mercury*, 23 May 1861, 4; "Buffington Family." In 1862, the Confederates reportedly burned the lumber in Volusia County to keep it out of Union hands.

observed, "Regardless of pleas for Floridians to unite behind secession, the first fall of the war demonstrated that the state's citizens could not forget their past political disagreements.... Milton used his position to attack not only his political opponents...but also those [military officers] who disagreed with his judgment." As a result, "[d]uring the fall [1861], Milton quarreled with every colonel in the state, except [James Patton] Anderson, whom he revered."[17]

The Confederate commander in Florida should have served as a buffer between Milton and the state's military, but Brig. Gen. John Breckinridge Grayson was in no physical condition to halt Milton's shenanigans. A Kentuckian, a graduate of West Point, and a veteran of both the Second Seminole and Mexican Wars, Grayson had been appointed by Richmond to oversee the District of Middle and Eastern Florida. The Kentuckian, however, was in the final stages of tuberculosis. Despite his debilitating illness, Grayson tried to fulfill his duties. He began a tour of his new department, but his subordinates reported to Pres. Jefferson Davis that Grayson had been "rendered *non compos mentis* by his disease." One of Grayson's last reports cautioned that Florida "will become a Yankee Province unless measures for her relief are promptly made." Perhaps assuming that the Kentuckian was delusional, the Confederate politicians in Virginia ignored his warning.[18]

Grayson died on 21 October 1861, and the Confederate government named Brig. Gen. James Heyward Trapier to replace Grayson. After graduating from West Point, the native of Georgetown, South Carolina, received an appointment to the coveted engineering corps "because of his high class standing." Almost immediately after assuming his post in Florida, he and Milton "crossed swords" on the necessity maintaining a cavalry force in the state. Trapier averred that horse soldiers would be needed if the Yankees gained a foothold in Florida while Milton worried about the costs to the state of feeding and equipping mounted troops.[19]

Whether correctly or incorrectly, the state's citizens quickly concluded that Trapier lacked the courage to fight, and a "state convention introduced a resolution requesting [requiring] that the Rebel government either force

---

[17] Sheppard, *Noble Daring*, 20–21; Murfree, "Rebel Sovereigns," 59–61.

[18] W. C. Davis, "John Breckinridge Grayson," in *Confederate General*, 3:28–29.

[19] T. Jones, "James Heyward Trapier," in W. C. Davis, ed., *Confederate General*, 6:58–59; Sheppard, *Noble Daring*, 31–32.

Trapier to remain with his troops 'at some threatened and exposed point,' or replace him." Feeling "greatly insulted," the general asked to be relieved of duty. The government in Richmond speedily complied with his request. (Trapier eventually served with the Southern armies in Mississippi and at Charleston but ended the conflict at his home in Georgetown.[20])

FERNANDINA

Fernandina, with a population of about fourteen hundred, seemed on the cusp of becoming a boomtown in early 1861. An article on Southern ports reported,

> The first important seaport after leaving Savannah is Fernandina, near the entrance of the St. Mary's River, the boundary between Cumberland and Amelia Islands, with fourteen feet of water at the bar. Fernandina is connected by a railway, one hundred and thirty-five miles in length, running across the state, with Cedar Keys, on the Gulf of Mexico, and thus an important commercial point.

The United States "Blockading Strategy Board," composed of high-ranking US naval officers and political officials in the Lincoln Administration, had reached the same conclusion. They had "recently identified Fernandina, Florida, as a desirable depot for provisions, coal, and a harbor or refuge for the [blockading] squadron. The inlet had twenty feet of water at high tide…[and the] area had no large populace and its isolation by the surrounding marshes would simplify the defense of the town." The Confederacy had reached basically the same conclusion, and that put the Amelia Island city squarely in the crosshairs of the two competing nations.[21]

Only one soldier and two workmen had been stationed at Fort Clinch, the unfinished bastion on the northern end of the island, and they hastily departed when Florida voted to leave the Union. Col. William Scott Dilworth, with several companies of his 3rd Florida Regiment, and Lt. Col. Daniel Pope Holland, with four independent companies, quickly moved in to fill the void. A comedy of errors ensued that would hamper the Rebel operations at Fernandina through most of 1861. As Don Hillhouse observed, "Fort

---

[20] Ibid.

[21] "Southern Harbors of the United States," *Amherst [NH] Farmer's Cabinet*, 20 September 1861, 2; Browning Jr., *Success Is All That Was Expected*, 20–21; Edling, "Fernandina Beach, Florida."

Clinch, the masonry fortification guarding the main approach to Fernandina through Cumberland Sound [what is today known as the Intracoastal Waterway], was still unfinished after more than a decade of construction and not structurally ready for emplacement of guns." The Rebel leaders concluded that the main attack would come from the Atlantic and began constructing sand and palmetto log bunkers along Amelia Island's north and northeastern shoreline. The guns, however, were "badly mounted, and would not stand continued firing." The problem, as an officer in the 3rd Florida observed, was: "[W]e have not a military company at this post [Amelia Island] capable of service as artillerists. Few of our volunteers have ever seen anything larger than a musket before coming to this station." After several more unsuccessful attempts by the Floridians to install the cannons, Gen. Robert E. Lee, in November, sent a naval officer who oversaw the construction of the gun emplacements and taught the Floridians the rudiments of firing the artillery.[22]

By that time, however, the war had already reached Nassau County. During the first week of August, the Rebel privateer *Jefferson Davis*, on her final voyage, encountered the US Sloop of War *Jamestown* in Cumberland Sound. The *Davis* had been escorting a captured prize, the *Alvarado*, to St. Augustine when the *Jamestown* intercepted the sea wolf. The *Davis* successfully took cover in the mouth of the St. Marys River, but the *Alvarado*, manned by a picked crew, ran aground a mile from the shore opposite the Nassau County city. The citizens of Fernandina responded to the crisis with alacrity. A participant in the "fight" stated, "Soon the drum was beating 'to arms' and in a short time the 'Fernandina Volunteers,' 'Island City Guards,' and also private citizens were armed and on their way to the scene of the action." In addition, "[a] company from the post at Fort Clinch, with a 6-pounder, was also dispatched to the beach...[which] made up the whole of our...armament." As soon as the Nassau County "warriors" reached the locale, they immediately opened fire on the *Jamestown* and her crew.[23]

The privateer's crew, on the *Alvarado,* lingered aboard their prize long enough to salvage what they could from the captured vessel while the Union warship fired at them. The Rebels managed to rescue about $10,000 worth

---

[22] Hillhouse, *Heavy Artillery & Light Infantry*, 4–7; Sheppard, *Noble Daring*, 28–32; *ORA*, ser. 1, vol. 6, 312.

[23] Schmidt, *Florida's East Coast*, 32–41; US Navy War Department, *Official Records of the Union and Confederate Navies in the War of the Rebellion* [hereafter *ORN*], ser. 1, vol. 6, 59.

of the *Alvarado*'s cargo before abandoning the beached vessel. A crew from the *Jamestown*, in three launches (described by a Fernandina resident as "row boats") soon arrived and tried to free the beached vessel under a hail of musket and cannon fire. Finding it nearly impossible to recover anything of value from the *Alvarado*, the Union sailors set fire to the vessel and returned to the *Jamestown*. As soon as the launches returned, the blockader's crew gave up the fight and retreated to the open water.[24]

Louis Coxetter, captain of the *Jefferson Davis*, set the *Alvarado*'s master, S. J. Whiting, his wife, and a Black cook ashore at Fernandina before sailing for St. Augustine. The three Northerners likely feared the worst—ostracism, torture, imprisonment, or death—at the hands of the "bloodthirsty" Southerners. The ladies of the town, however, led by "Mrs.-ex-Senator Yulee started a subscription for the purpose of supplying Mrs. Whiting with suitable apparel for her journey [to the North]." The Whitings were wined and dined by Fernandina society, who also provided their "guests" with free passage by rail to Boston. When Capt. Whiting reached Massachusetts, he made certain that the local newspapers printed an account of the good treatment that his "family" received at Fernandina.[25]

CEDAR KEYS

Compared to Northeast Florida, with its sizeable towns and prosperous plantations, the state's Gulf Coast region must have seemed little more than a barren wilderness in the 1860s. A contemporary editor described that part of the state as a "dim and gloomy" region covered with "primeval and almost impenetrable forests." A modern historian concurred, labeling the area "a desolate shore." Levy County, according to the 1860 Census, had a population of 1,781, with a significant percentage of that total being Black slaves. The county's largest town seems to have been Cedar Keys, with a population of fewer than two hundred residents. The little port had a somewhat checkered history, having at various times served as a haven for gunrunners, smugglers, and even runaway slaves. By 1860, however, Cedar Keys had assumed an aura of respectability, shipping the region's chief exports—turpentine, fish, turtle meat, lumber, and cattle—to the larger ports along the Gulf Coast and

---

[24] Ibid. "Detailed Account of the Burning of the Alvarado," *Baltimore [MD] Sun*, 13 September 1861, 1.

[25] "Rebel Treatment of Prisoners," *New York Commercial Advertiser*, 23 August 1861, 2.

Florida Keys. The articles on Southern ports, widely published in Northern journals, described the Levy County village as, "The first [Florida] port on the Gulf of Mexico of commercial value is Cedar Keys, situated ten miles south of the denouchment [mouth] of the Suwannee River.... The entrances of Cedar Keys' harbor are narrow; the best has a depth of only eleven feet over the bar."[26]

The coastal hamlet might have slumbered in relative tranquility throughout the war, but Sen. David Levy Yulee had selected the village as the western terminus of his Florida Railroad. At the Levy County seaport, products and goods from Galveston, Texas, to Apalachicola, Florida, could be loaded on trains for transport to the Atlantic shore without the costly, dangerous voyage around the Florida peninsula. (Though largely forgotten in modern times, "[Yulee's] intent was to bypass the hazardous Florida Straits, an area where ship sinking annual los[s]es reached millions of dollars.") Local historian Charles C. Fishburne Jr. noted, "Less than a month prior to the outbreak of hostilities, the final section of the Florida Railroad was inspected and certified by the State Engineer John Bradford. It was then, with all trestles and turntables and rails to the wharf completed that the first train came through the terminal on Way Key."[27]

With Cedar Keys' increased importance, Gov. Perry felt compelled to protect the new transportation hub. In early July 1861, two small companies of the 4th Florida Infantry, commanded by Maj. Whit Smith, arrived at Cedar Keys. The politicians at Tallahassee apparently harbored serious reservations regarding Smith's competency and his men's military fitness. (The questions regarding Smith appear to have been well founded; the regimental surgeon described his commanding officer as "another drunken no-Count-man.") Those companies were, however, the only state force available to defend the Levy County port.[28]

---

[26] Cedar Keys and Cedar Key have been used interchangeably since the 1700s. I have used "Cedar Keys" in the text but do not change it in quotes and book titles or use the annoying "[*sic*]" to designate the "Cedar Key" usage. See McCarty, *Cedar Key, Florida*, 27–29; "Southern Harbors of the United States," *Amherst [NH] Farmer's Cabinet*, 20 September 1861, 2; P. Taylor, *Discovering the Civil War in Florida*, 151–54; Buker, *Blockaders, Refugees, & Contrabands*, 6; R. Taylor, *Rebel Storehouse*, 18.

[27] Fishburne Jr., *The Cedar Keys in the 19th Century*, 59–61; "Preserving Florida's Railroad History."

[28] Sheppard, *Noble Daring*, 33.

Maj. Smith and his men quickly got a chance to prove their military fitness. On 3 July 1861, Rebel lookouts spotted four vessels, flying the Stars and Stripes, becalmed on the Gulf, two or three miles west of Cedar Keys. Smith decided to take the enemy ships, but he faced two problems. He had neither ships nor experienced sailors in his companies. He solved the first deficiency by commandeering a steamer named the *Madison*. The vessel was a converted ferryboat, which Smith outfitted with a single six-pound brass cannon. Shortly thereafter, Major Smith discovered that one of his men—Lt. W. D. Burtchaell—had some experience as a saltwater seaman.[29]

Burtchaell, who seems to have taken command of the operation, related,

> As we [Maj. Smith, Burtchaell, and several volunteers] were not prepared for a sea fight and as I was the only soldier on board who had ever been on salt water, we prudently sailed for the two smaller [Union] vessels to the northwest of the mouth of the harbor. When we neared them we discovered that the men on board were soldiers [sailors] of the U. S. Navy. We took both vessels in tow and removed the Yankees to our steamer [and] as we were desirous of learning what and who were aboard the others... [W]e gave each prisoner a cup of grog and soon had all the information...[we] needed.[30]

While Maj. Smith's men were dealing with the first two ships, the two remaining vessels attempted to slip away, but they failed to make much headway on the dead calm sea. Burtchaell continued his narrative, relating that as soon as the first two boats and prisoners were secured,

> we started to take the two [remaining] vessels at sea....When we were sufficiently near them a blank cartridge from the six-pound gun was fired to stop them, but as no heed was taken, a solid shot was sent across their bows, and as they continued on their way, a shot was sent that [was intended to] hit the largest [sail]. It passed between the masts. We were out of ammunition except for one load of powder. This, along with spikes and pieces of china were put in the gun...Major Smith [then] ordered me to take six men in a small boat, to board the vessels and demand their surrender....

---

[29] *ORN*, ser. 1, vol. 6, 301; Collins, "The Saga of the Steamboat *Madison*," 16; Burtchaell, "Four Vessels," 13–17.

[30] Burtchaell, "Four Vessels."

> He [the Union ship's commander] said they were sorry that they did not have a gun to return the salute; and in reply, I told him that "[everyone knew that] a U. S. Navy officer would not surrender without a fight, if he could help it." With his pride somewhat assuaged, Lt. George L. Selder [actually, Selden] bowed to circumstances and acknowledged defeat.[31]

Florida newspapers reported the capture of the four ships and their cargoes: "[T]he brig *Three Brothers*, loaded with brick, the schooner *Olive Branch*, loaded with turpentine, the schooner *Fanny*, loaded with railroad iron, and the schooner *Rasilica*, [also] loaded with turpentine." Fifteen Confederate prisoners and a Black man [loyal to the Confederacy] had been found on the enemy vessels. They "rejoiced at their unexpected escape from confinement in the Fort at Key West." All of the former captives stated their intention to return to New Orleans, their original port of call.[32]

Burtchaell received the honor of escorting the prisoners to Tallahassee where he received a hero's welcome. The enlisted sailors were housed in the town jail, but Lt. Selden, at Burtchaell's insistence, spent his time in Tallahassee as a guest at the Gamble Plantation. Selden was related to the Gambles, but Burtchaell's act of kindness backfired. Northern newspapers immediately branded the Virginia-born Selden a turncoat—a stain on his reputation he apparently never lived down. Seizing on Selden's posh confinement, as opposed to the harsher incarceration of his men, a New York journal flatly stated, "Lieut. Selden, who is a Virginian lost the prizes in his charge purposely." Burtchaell categorically denied any collusion by Selden, describing the mariner as a brave man and loyal bluecoat.[33]

The conventional wisdom that "lightning does not strike twice in the same place" did not apply to Lt. Burtchaell. A month later, a watchman spotted a ship, flying the flag of Bermuda, passing within sight of Cedar Keys. Burtchaell "determined to take [it] if possible." The young lieutenant put five men in a sloop and began his pursuit. He kept four of his men below deck, "leaving only the man at the rudder besides myself to be seen from the [enemy] ship. After a fifteen-hour race, the Confederates pulled beside the vessel

---

[31] Ibid.

[32] Ibid.; "Recapture of United States Prizes at Cedar Key," *Charleston [SC] Courier*, 16 July 1861, 3.

[33] Burtchaell, "Four Vessels;" "Interesting from Washington," *Albany [NY] Evening Journal*, 7 August 1861, 2.

and raised the Stars and Bars. At the same instant, the four men sprang from their place of concealment, leveling their muskets at the startled sailors." The sight of the armed Rebels took all the fight out of the enemy crew, and Burtchaell and two men climbed aboard the vessel. Despite protests that the men hailed from Bermuda, Burtchaell soon discovered that "this vessel was engaged in sea and green turtle fishing to feed the United States troops at Key West. We found thirty-six turtles weighing from twenty-five to six hundred pounds." Maj. Smith's men forwarded the ship to Mobile. Much to his relief, Burtchaell soon received orders to "the seat of the war" in Virginia. There, the "saltwater sailor" became a proud member of Brig. Gen. E. A. Perry's Florida Brigade.[34]

TAMPA AND SOUTH FLORIDA

On the morning of 13 January 1861, the stagecoach from Gainesville arrived at Tampa, bringing news that the state had voted to leave the Union. Many Tampa citizens greeted the news of secession with wild rejoicing. "Despite the fact that the day [the news reached the city] was Sunday," a local historian related, "young men rushed to the fort [Fort Brooke] and fired the cannons again and again. Members of the Tampa Brass Band hurriedly donned their uniforms and paraded up and down the streets [playing patriotic music]. And that evening, minsters prayed that Florida's leaders would be given Divine guidance in this hour of peril." The public outpouring of joy seems to have been heartfelt, but a significant minority in South Florida either supported the Union or simply wanted to be left alone.[35]

The City of Tampa had grown up around Fort Brooke, a bastion established in 1824 to protect the Bay region during the various Seminole Wars. Naturally, Confederate troops occupied the outpost during the opening days of the Civil War. In late August 1861, Maj. Wylde Lyde Latham Bowen, with companies D and E of the 4th Florida Infantry, took command at Tampa Bay. Bowen, a graduate of Mossy Creek Baptist College (modern Carson–Newman University) had recently moved to Florida to read law in the Lake City office of Whit Smith.[36]

---

[34] Burtchaell, "Capture of Vessel Off of Cedar Keys," 18–20.

[35] Grismer, *Tampa*, 137–38; C. Brown Jr., *Florida's Peace River Frontier*, 142; Ivey, "John T. Lesley," 11–14.

[36] Waters, "Tampa's Forgotten Defenders," 3–6; Sheppard, *Noble Daring*, 22–23.

Initially, the Union's investment of South Florida had more holes in it than a fishing seine, and Bowen moved promptly to break the enemy embargo. To put some teeth in the blockade, the Yankees had established a battery at Egmont Key, in the mouth of Tampa Bay. Despite the Union's attempt to shut off the arrival of supplies, local ship captains such as James McKay Sr., Frederick Tresca, and John W. Curry slipped in and out of the bay, almost at will. Blockade running proved too lucrative a trade to quit. As an early historian observed, "It was a fine business—when the ships were not captured. Huge profits could be made on both ends of the hazardous journey through the Union blockade."[37]

In early July, Maj. Bowen decided to take the offensive. First, he obtained the loan of the steamboat *Madison* from Maj. Smith at Cedar Keys. With the shallow draft vessel, he swooped down on a pair of Union ships fishing in Tampa Bay. A Southern newspaper reported that the steamer "fully armed and manned by two companies stationed there, went out to reconnoiter...and succeeded in capturing two schooners." A couple of weeks later, with the blockading squadron otherwise occupied, Bowen's men swept down on Egmont Key and "removed the oil and [light] fixtures from [lighthouse's] signal tower." A member of the local militia later reported: "No light is made [has been seen] since...our forces has stripped the lighthouse & fortifications lately made on the West end of Eggmont island [Egmont Key]."[38]

In late November, using Capt. James McKay's blockade-runner, *Scottish Chief*, Bowen "went in search of the fishing tribe. Armed with a six-pounder, he soon secured [captured] a fine fishing smack and brought her into Tampa Bay.... [Using the vessel just taken] he captured twelve sail[boats], nine smacks, three schooners, and [has] effectively broken up the infamous [Yankee fishing] traffic." Bowen and his troops also captured sixty-eight men, described as "some Yankees, some Spaniards, some Portuguese, and some Key Westers."[39]

---

[37] Grismer, *Tampa*, 139–41; "Egmont Key Occupied by Union Troops," *Washington [DC] Daily National Intelligencer*, 14 April 1861.

[38] "Confederate Coast Observations in South Florida," *Charleston [SC] Courier*, 15 December 1861, 1; "Blockade of Tampa Bay," *Richmond [VA] Whig*, 9 July 1861, 2; Shaw Point letter [no writer noted], Heaton family, 28 August 1861, photocopy in Zack C. Waters collection, Rome, GA.

[39] "Tampa," *Macon [GA] Telegraph*, 11 October 1861, 4; "State Items," *Hartford [CT] Courant*, 18 December 1861, 2.

Not all the action in South Florida, in 1861, occurred on the bay or in the Gulf. Union troops descended on Shaw Point [in modern Manatee County], demanding information on some buried artillery pieces. The frightened villagers had no idea what they were talking about, and the bluecoats eventually returned to Key West. In October, a Savannah journal noted, "A posse of Capt. J[acob] Summerlin's Cow Boys last week made a descent on a body of Lincolnites who were engaged in cutting timber near Fort Myers for…a man named [Lyman] Stickney who had a contract with United States authorities at Key West, and succeeded in capturing four of them." (Both Summerlin and Stickney would play significant roles in Florida during the war years.) Stickney, described by a modern historian as "a professional opportunist," had been living at Key West when he saw a chance to make some money by cutting timber, which could be converted into fuel for the blockading fleet. Of course, Stickney would never dirty his hands with physical labor, nor put himself in danger, so it was his hirelings who fell into Summerlin's hands. Two of the captured lumbermen were French nationals, trying to make enough money to return to their homeland. The Cracker cowman treated them with kindness, and even sent them to Savannah, Georgia, where they were more likely to find transport to Europe. The other two woodcutters, both residents of Key West, "were placed in confinement."[40]

The age-old process of demonizing the enemy had already begun in both Northern and Southern newspapers, but the Union characterizations of the South Florida Rebels seemed particularly vile. The *New York Tribune*, for example, reported,

> Trustworthy information has been received that the rebels in South Florida, whose headquarters are in Tampa are organized into bands of pirates and robbers. Those predatory outlaws, composed of African slavers, negro stealers, and scoundrels generally, more sanguinary and lawless than Billy Bowlegs and his savage clan, violently tear Union men from their homes and appropriate to themselves

[40] "Letter from Tampa," *New Orleans Times-Picayune*, 27 October 1861, 2; "Jacob Summerlin: The Cowman Who Was King of Crackers"; Coles, "Far from Fields of Glory," 9; Shaw Point letter; see generally, Akerman and Akerman, *Jacob Summerlin: King of the Crackers*.

> property wherever it is found. They have reduced rapine to an ethical science.[41]

Such vilification, of course, helped to make any atrocity practiced against that enemy seem justifiable.

[41] "From Key West," *New York Tribune*, 22 November 1861, 6.

Chapter 2

"A System of Annoyances"

(January–July 1862)

Situation in Early 1862

Initially, Confederate sympathizers considered Pres. Abraham Lincoln's widely publicized plan for a blockade of Southern ports a pipe dream, and Gen. Winfield Scott's "Anaconda Plan" seemed even more ludicrous. After all, in April 1861, the US Navy had fewer than fifty ships in active service, and most of those vessels were assigned to foreign stations. But, even if the enemy somehow managed to put an effective embargo in place, many in the new nation felt they had "an ace in the hole." Should the US attempt to shut off imports and exports, Great Britain, Europe's most powerful nation, would intervene on the South's behalf. After all, they reasoned, "in 1859, England had imported 78 percent of the South's cotton crop…which provided four to five million Englishmen with jobs." And, as long as they could buy arms and ammunition from Europe, Southerners felt confident that their Rebel soldiers could beat the Yankees in a standup fight.[1]

Nothing, however, went according to that carefully crafted script. "[B]y the end of the first year [of the war] 52 [Union] vessels had been built and 136 had been purchased; and, Great Britain had declared its neutrality regarding the American 'difficulties'." Florida's citizens felt the pinch almost immediately. "In the early months of the War Between the States," historian Mary B. Graff explained,

> there was no demand for supplies, since the theater of the war was in Virginia; furthermore, the blockade was ineffectual and goods were frequently brought into Florida from the West Indies. As the months progressed, however, stocks of all kinds gradually diminished, and by spring 1862, salt, hardware, groceries, shoes, dry goods, medicines, and small wares became scarcer and [more]

---

[1] Wise, *Lifeline of the Confederacy*, 10–13; Page, *Ship versus Shore*, 11–16; Buker, *Blockaders, Refugees, & Contrabands*,1–3.

> expensive. Confederate money depreciated in value and all available goods brought fabulous prices.

The Union's stranglehold on importation of foreign goods would continue to tighten as the war dragged on toward its conclusion.[2]

CEDAR KEYS

The first military action in Florida, in 1862, occurred at Cedar Keys. The Federal Navy had been blockading the Levy County coast since mid-1861, but the initial embargo proved largely ineffective. The local sailors knew the Levy County waterways better than the blockaders, and small Southern vessels reportedly slipped through Union cordon with considerable frequency. Cedar Keys' position as the western terminus of the Florida Railroad increased the town's value as a military target, and the Yankees wanted a chance to avenge the well-publicized capture of Union ships by Maj. Whit Smith's sea-going soldiers. As a result, the Federals made an early attack on the island town a top priority.[3]

About nine o'clock on the morning of 16 January 1862, the Union warships *Hatteras* and *Florida* anchored just outside the bar at Cedar Keys. Com. George F. Emmons, skipper of the *Hatteras*, ordered his men to fire three warning shots over the town, but the Rebels failed to respond. Yankee sailors quickly entered the port city and destroyed all military targets.: "I have been entirely successful…," Com. Emmons wrote,

> capturing or destroying all the public property here, including a battery of two long eighteens [eighteen-pound cannons] in position on the east side of Sea Horse Key, with their carriages and some ammunition and barracks, a six-pound field piece on Depot Key, with the railroad depot and wharf, several [train] cars,

---

[2] Wise, *Lifeline of the Confederacy*, 10–13; Page, *Ship versus Shore*, 11–16; Buker, *Blockaders, Refugees, & Contrabands*, 1–3; quote is from Graff, *Mandarin on the St. Johns*, 35–36.

[3] Cox, "Union Attack on Cedar Key—Cedar Key, Florida"; "The Capture of Cedar Keys," *New York Herald*, 26 January 1862, 1; Proctor, *FL100*, 16 January 1862.

telegraph office, and turpentine storehouse, besides four schooners and three sloops, one ferry scow, sailboat, and launch.[4]

Com. Emmons and his men also captured fourteen Southern soldiers, members of the 4th Florida Infantry, and several women. (Most of the troops previously stationed in the Levy County port city had been withdrawn to strengthen the Amelia Island defenses.) Gen. Trapier, however, had bowed to pleas of the citizens of Cedar Keys, by sending a tiny force to protect the residents against "bands of marauders." He explained: "[C]itizens from that community [Cedar Keys]...feared that certain persons who had been arrested as traitors, and released...[would] avail themselves of the opportunity...to wreak their vengeance upon their accusers." When the *Hatteras* fired her warning shots, the Confederate soldiers and some local ladies attempted to flee to the mainland using an old ferryboat, "but they found when they reached mid-channel that their poles were too short to reach the bottom. They were at the mercy of the tide, by which they were swept out and fell easy victims into the hands of the enemy." Rather than being taken to Key West—or worse yet, Fort Jefferson, in the Dry Tortugas—the Union sailors quickly freed their prisoners when they discovered four Rebel soldiers had measles. Afraid that the infected men would spread the disease throughout the Federal fleet, the bluecoats paroled all of the Rebels and banished them from the town.[5]

The residents of the little port city knew they had been fortunate. A letter from the town averred, "[T]he citizens of Cedar Keys were not molested, and [private] property was not taken," but the raid "demonstrated that they [the Federals] could deny the Confederate authorities reliable use of the port and rail terminus at their pleasure." A New York journal went further in its assessment of the naval operation. "The Florida Railroad, running as it does almost directly across the peninsula, afforded a splendid opportunity for conveying goods and troops across the country. This, by the taking of Cedar

---

[4] "The Attack on Cedar Key—Cedar Key, Florida"; *ORN*, ser. 1, vol. 17, 48–49; Proctor, *FL100*, 16 January 1862. The *Hatteras* would be sunk in the Western Gulf the following year by the CSA "pirate" *Alabama*, commanded by John N. Maffitt.

[5] *ORN*, ser. 1, vol. 17, 48–49; *ORA*, ser. 1, vol. 6, 75–76; "From Cedar Keys, Florida," *Lancaster [SC] Ledger*, 29 January 1862, 1; Proctor, *FL100*, 16 January 1862.

Keys, has been effectually stopped, and a long and dangerous trip around the coast is the consequence."[6]

Several Southern newspapers described a second encounter, in early March, near Cedar Keys, but it may have been an example of pro-Confederate journalists exaggerating a very minor incident into a much-needed Confederate "victory." The articles related that Union sailors raided the Levy County mainland, "committing a variety of depredations." The correspondent continued,

> A detachment of our [Southern] troops were sent to cut off their retreat but failing in this, they immediately opened fire upon the Yankees as they were making the best speed they could toward their vessels. The enemy returned four shots [from the ship's cannons], and quit firing. Our men kept up a brisk fire until they [the Union soldiers] were beyond our reach, which resulted in the loss of...[an] estimated...ten or fifteen [Federal raiders.][7]

NEW SMYRNA

Almost as soon as Florida voted to secede, blockade running began in coastal Volusia County. Small steamers and sloops carried cotton and other goods to the Bahamas, just 350 miles to the southeast. British commercial ships often anchored at Nassau and paid top dollar for Rebel cotton. The enterprise was not easy money because the blockade-runners could face considerable difficulties in reaching New Smyrna. After dodging the blockading vessels, they had to cross the shallow bar at Mosquito Inlet [modern Ponce de Leon Inlet], and then meander along the narrow, constantly shifting channel to reach the tiny hamlet. But the reward made the risk worth the effort.[8]

On 17 January 1862 (the day after the Union attack on Cedar Keys), Gen. Robert E. Lee, commanding the coastal defenses in Florida, Georgia, and South Carolina, ordered Gen. Trapier to have "two moderate-sized" artillery pieces implanted on the mainland to protect the entrance to Mosquito

---

[6] Cox, "The Union Attack on Cedar Key – Cedar Key, Florida"; "The Capture of Cedar Keys; Official Account of the Taking of Cedar Keys by the Enemy," *New Orleans Daily True Delta*, 1 February 1862, 2; Fishburne Jr., *The Cedar Keys in the 19th Century*, 61.

[7] "From Cedar Keys," *Augusta [GA] Daily Constitutionalist*, 13 March 1862, 2.

[8] Bowden, "Volusia's Bloodiest Civil War Battles"; Sweett, *New Smyrna*, n.p.; "Our History [The Bahamas]."

Inlet. Lee warned Trapier that the cargoes to be landed there were "so vitally important that no precaution [to protect it] should be omitted." This action represented the culmination of an operation that began in England the previous year, and repercussions from this enterprise would bring combat and bloodshed to the New Smyrna area for the next two years.[9]

In April 1861, Pres. Jefferson Davis had sent Maj. Caleb Huse to England to purchase military supplies for the new Confederate nation. Born in Massachusetts, Huse had become close friends with Davis at West Point. After serving a few years in the US Army, Huse retired and assumed a teaching position at the University of Alabama. Huse and Davis both recognized that the South, with its limited manufacturing facilities, could never produce weapons comparable to those supplied to the typical Yankee soldier. At the Davis's behest, Huse sailed to England and purchased various military materials, including more than forty thousand Enfield rifles for the new nation. However, with the vastly improving Union blockading squadrons, getting the firearms to the Southern mainland proved difficult.[10]

The first load of English arms and equipment, aboard the *Bermuda*, ran through the blockade into Savannah. Federal vessels routinely patrolled the mouth of the Savannah River, but the blockade-runner sailed straight up the waterway without a hint of interference. The cargo included "at least eighteen rifled field pieces [cannons], four heavy seacoast guns, 6,500 Enfield rifles, and 20,000 cartridges." The *Bermuda*'s load also contained 24,000 blankets, 50,000 pairs of shoes, and 200 Enfield carbines. Though luck had smiled on the Confederacy once, the cargo was too valuable to trust to blind fortune twice.[11]

Huse dispatched the next shipment, including 20,450 British-made rifles, aboard a heavy ship called the *Gladiator.* As she neared the Florida coastline, Union warships spotted the blockade-runner and chased the vessel. Rather than surrendering to the Union blockaders, the *Gladiator* took refuge in the Bahamas. From there, the *Gladiator* hoped to make a run for a Southern port; but those plans were canceled when Union cruisers began routinely patrolling the waters around the Bahamas in an attempt to keep the weapons from reaching the Southern armies. A naval historian observed, "Any attempt

[9] Wise, *Lifeline of the Confederacy*, 49–51; *ORA*, ser. 1, vol. 6, 370.

[10] Wise, *Lifeline of the Confederacy*, 49–51.

[11] Ibid.

by the heavily laden *Gladiator* to outrun the armed steamers was considered fruitless, and for the moment the Confederates were stalemated."[12]

To solve that problem, Gen. Lee turned for advice to Lt. John Newland Maffitt. Born at sea, between Ireland and the United States, and raised in North Carolina, Maffitt had more than fifteen years of prewar naval service, primarily in the American South with US Coastal Survey units. In early 1862, he had received an appointment as a special aide to Robert E. Lee. Maffitt made two suggestions that the Virginian wisely accepted. He recommended running the military cargo into Mosquito Inlet, near New Smyrna. Maffitt also counseled using smaller, faster blockade-runners that would allow the Rebels to outrun the more cumbersome Yankee ships and take refuge in places the bigger enemy vessels could not easily enter. Two vessels, the *Kate* and the *Cecile*, made the short run to Florida's east coast without encountering any enemy interference. "Once [un]loaded, the guns were placed into wagons and carried to [Enterprise on] the St. Johns River, where [Confederate] steamers transported the supplies to the railroad at Jacksonville." From there, Lee dispersed the munitions to the Rebel armies in Tennessee, South Carolina, and Virginia.[13]

AMELIA ISLAND

Robert E. Lee's time in Northern Florida had convinced him of the impossibility of protecting state with conventional Confederate soldiers. With the military failures in Tennessee—and with the Old Dominion under constant Union pressure—calls for Florida troops seemed to arrive at Tallahassee almost weekly. On 24 February, Lee bluntly stated the obvious: all Florida troops must be sent to Tennessee or Virginia, leaving only those fighters "necessary to defend [the] Apalachicola River, by which the enemy gunboats may penetrate far into the State of Georgia." While it was a correct assessment, from a strictly military perspective, Lee's directive outraged most Floridians.[14]

A true Southern nationalist, Gov. John Milton fumed at the abandonment of his state by the Davis Administration, but he did what he could to keep the state's White residents encouraged. He began by establishing a short-lived militia, whose primary purpose was to prevent slave rebellions. Such

[12] Ibid.; see generally, Sword, *Firepower from Abroad*, 17–28.

[13] Wise, *Lifeline of the Confederacy*, 57–59; see generally, Shingleton, *High Seas Confederate*. Enterprise was the county seat of Volusia County in the 1860s.

[14] *ORA*, ser. 1, vol. 6, 403; Reiger, "Florida after Secession," 130–32.

rag-tag, poorly armed units, however, could not be expected to do much against trained and well-equipped Federal soldiers, and those weak measures offered the average citizen little reason for optimism. H. H. Hoeg, Jacksonville's mayor and a native of New York, seemed to speak for many Floridians when he issued the following declaration: "Inasmuch as all Confederate troops, arms, and munitions of war upon the St. Johns River and in the east and south Florida generally are ordered away, and that the east and south are to be abandoned [by the Confederacy], it is useless to attempt a defense of the city [Jacksonville]...and therefore, upon approach of the enemy, it should be surrendered." Hoeg was a died-in-the-wool Unionist with a personal agenda, but his reasoning must have made sense to many native Floridians.[15]

The residents of the southernmost state did not have to wait long before the Federals arrived. On 3 March, a Federal armada, commanded by Flag Officer Samuel F. Du Pont, sailed into Cumberland Sound, intent upon capturing Fernandina. His flotilla consisted of about thirty ships and included Gen. Horatio G. Wright with four infantry regiments, two artillery units, and various volunteers. Du Pont knew precisely what he faced as Florida Unionists and escaped slaves had been keeping him apprised of Confederate activities in Northeast Florida.[16]

Luckily for Florida, most of the Confederate forces had departed the area before the Unionists arrived at Amelia Island. The final military contingent, a part of the 1st Florida Special Battalion, had been digging entrenchments and emplacing cannons near Ft. Clinch when the orders arrived to leave immediately. George H. Dorman, a member of the unit, recalled, "We received orders to spike the guns and evacuate the place [Fernandina]. We left the island that night.... [A]ll hands walked across a long railroad trestle in the dark, and luckily no one fell in."[17]

Many civilians had also decided to depart Fernandina. A Federal naval officer described the chaotic scene that greeted the Yankees when they neared Amelia Island:

---

[15] Reiger, "Florida after Secession," 130–32; Schafer, *Thunder on the River*, 51–52.

[16] Schafer, *Thunder on the River*, 47; *ORN*, ser. 1, vol. 12, 46–47.

[17] Schafer, *Thunder on the River*, 48; Hillhouse, *Heavy Artillery & Light Infantry*, 24; "Glorious News from the Navy," *Washington [PA] Reporter*, 13 March 1862, 2; "Our Occupation of Florida," *Jamestown [NY] Journal*, 11 April 1862, 1.

> As the fleet approached the Fort [Clinch], a train of cars was observed leaving Fernandina, and as the tracks run some miles along the shore of the [S]ound, Com. Du Pont sent one of the gunboats in pursuit of it. An exciting race took place. The steamer threw shells at the flying train and some of them falling in such proximity that some of the fleeing rebels jumped from it [the train] and took to the bushes. Among the latter is said to have been the late [former] Senator Yulee, of Fla. accompanied by his servant.[18]

The train managed to outdistance the gunboat and escaped. Another newspaper reported that the gunship *Ottawa* shelled the cars, but the captain of the *Darlington* [just captured by the bluecoats] refused to do so because of the presence of women and children on the train. Thus, in disorderly and panicked flight, the Confederate occupation of Amelia Island ended.[19]

The Federal troops quickly occupied Ft. Clinch, one of the obsolete, prewar brick bastions, and expressed amazement at the amount of war material left by the fleeing Confederates. They discovered "eight Rebel earthworks abandoned by the enemy. Twelve large guns fell into our possession including one immense rifled gun of 120-pound calibree [caliber]. Five of them were at Fort Clinch and others at the earthworks." A considerable number of wagons, plus ammunition, and camp equipment also fell into the Federals' hands, as well as the local excursion and mercantile vessel *Darlington.* The Union troops did not, however, confine their scavenging to military targets. One of the bluecoats reported, "The town was in the utmost confusion. The troops already landed were plundering in all directions and were ably assisted by sailors and marines from the fleet who seemed to have reduced the pillage to an exact science. Drunken men [soldiers, sailors, and marines] were seen everywhere, and the few inhabitants remaining, were waiting in their homes, terrified." The Unionists took particular pride that the occupation of Amelia Island occurred on the anniversary of the inauguration of Pres. Abraham Lincoln. One giddy reporter—in a groan-worthy attempt at humor—wrote, "The U. S. forces some weeks ago, drove a strong nail in the other side of

[18] *New York Times*, 11 March 1862.

[19] Schafer, *Thunder on the River*, 47–48; "Glorious News from the Navy"; "Our Occupation of Florida."

Florida at Cedar Key, and now Com. Du Pont has *clinched* [emphasis added] it on the other."[20]

Most of the Federal troops seemed less than impressed with their new duty station. A Union trooper described Fernandina as a "small place, of about two thousand inhabitants, near the northern end of Amelia Island." (The 1860 Census placed the town's population at 1,390, and since many of the White residents left on the last train, the actual number likely totaled between eight and nine hundred. An artist from *Frank Leslie's Illustrated Newspaper* characterized Fernandina as "a most dismal and mouldy-looking place. It is inhabited by a few Whites and more Blacks. But they seem to have little to do, and partake of the melancholy character of the place. The landing is very dilapidated—in a word, it conveys the idea of decay." In the Nassau County countryside, an equally miserable scene greeted the invaders. Another writer observed, "At every point along the South Atlantic coast…were abandoned plantations, deserted towns, wasted crops, stray cattle, and wandering slaves…. The people believed the Yankees would treat them [the White civilians] brutally, and left their homes in terror."[21]

Through that torrent of depressing news, Florida's pro-Confederate populace found a glimmer of hope. It all began with a blockade-runner named *Hard Times*. During the first year of the war, Fernandina had been a favorite destination for the fleet ships bringing necessities and luxury items into the Rebel states. Having been at sea for several days, the captain of the *Hard Times* did not know of the recent Federal occupation of Fernandina. As the little steamer approached Amelia Island, the skipper observed the Stars and Stripes floating over Fort Clinch, boldly sped past town, and proceeded straight into the St. Marys River. Loaded with a large and valuable cargo of guns and ammunition, the *Hard Times*'s captain managed to evade the

[20] "Commodore DuPont's Expedition on the Southern Coast," *Elkton [MD] Cecil Whig*, 15 March 1862, 3; [No headline], *Evansville [IN] Daily Journal*, 17 March 1862; Edling, "Fernandina Beach, Florida."

[21] "Old Fernandina, Florida," *Frank Leslie's Illustrated Newspaper*, 19 April 1862, 10; "Fernandina, Florida," *Montpelier [VT] Daily Green Mountain Freeman*, 12 March 1862, 2; "The Civil War Fifty Years Ago Today," *Lexington Leader*, 9 March 1912, 9.

pursuing Union vessel by taking refuge in a creek too small or shallow for the Yankee ship to enter.[22]

A dispatch for assistance quickly reached the nearby camp of the 1st Florida Confederate Cavalry at Jacksonville, regarding the *Hard Times*'s predicament and its valuable consignment. The unit's commander, Col. W. G. M. Davis, directed the companies of Captains Noble A. Hull and W. D. Clarke to proceed to the St. Marys to protect the stranded vessel and its cargo. Several citizen volunteers also joined the Rebel troops, making it a formidable force for that isolated region. A veteran of the ensuing skirmishes recalled,

> detachment of one hundred and fifty men were sent to the relief of the *Hard Times*, under command of Capt. N. A. Hull, and a little after sunrise...at Crozier's Bluff, the [Federal] gunboat was discovered going down the stream very rapidly...Capt. Hull then marched his command at a gallop four miles down the river to White Oak Bluff, getting ahead of the [Union] boat.

Hull and Clarke's units sprang their trap with "each man taking [cover behind] a tree, and with their Maynard rifles and double barreled shotguns, as the enemy gunboat got within sixty yards, the first of the ambush line opened, and the fire told with deadly effect upon the thickly crowded decks of the gunboat." According to a widely repeated story, a Union sailor in the ship's rigging spotted Capt. Clarke and yelled, "You d—d cowardly rebel." Clarke responded, "'You are a d—d liar,' as he pulled the trigger and 'settled his hash.'" A third article averred, "One...volunteer, a noted hunter and excellent marksman, fired five times, and each time selected his man—the ones with brass buttons on [officers]."[23]

The Federal steamer eventually broke free of the Rebels' trap, but their ordeal had not yet ended. The Florida troops continued pursuing the Federal ship along the south bank of the St. Marys, firing at their foes from a distance. Help arrived from three South Georgia home guard companies, commanded, it is believed, by Captains Thomas S. Hopkins and Enoch Douglas Hendry,

---

[22] "Attack on the Yankees near St. Mary's," *Fayetteville Carolina Observer*, 17 March 1862, 3; "The Affair on the St. Mary's," *Augusta [GA] Daily Constitutionalist*, 4 April 1862, 1. See also Waters, "Our Necessary Resort," 72–88.

[23] "Attack on the Yankees near St. Mary's"; "The Affair on the St. Mary's"; "Movement and Spirit of the War," *Richmond [VA] Enquirer*, 19 March 1862, 2; "More Arms," *Augusta [GA] Daily Constitutionalist*, 19 March 1862, 3; "Slaughter of the Enemy Near St. Mary's," *New Orleans Times-Picayune*, 23 March 1862, 5.

both of Ware County, and Capt. Alexander Lang, of Camden County. When the enemy ship came "within point blank range," the Georgia marksmen opened fire, pouring a devastating volley into the enemy sailors from the north shore. After that encounter, the gunboat moved out of range of the Rebels' rifles. Exact Union casualties on the expedition will likely never be known, but a Black man "who had...escaped from Amelia Island to the [Confederate] camp...states that he had to assist in burying 47 of the Yankees, and reports there were 16 wounded."[24]

Col. Davis ordered members of his regiment to guard the mainland and limit further incursions from Amelia Island. Initially, at least, they experienced some successes. In early April, Capt. William Footman's company captured two Federals they found heading into the Nassau County interior, along the railroad, using a handcar. A short time later, five men from the Ninth Maine appeared at the home of Judge O'Neal, and Footman's unit also took them prisoner. The graycoats killed one man who resisted, and Col. Davis sent the rest to Tallahassee as prisoners of war.[25]

St. Augustine

St. Augustine and Jacksonville both fell to the Union troops on the same day—11 May 1862. The Federals apparently expected a severe fight in their effort to wrest the Ancient City from the Rebels, so Gen. Thomas W. Sherman and Adm. Samuel Du Pont developed a complex plan of attack. Union ships would sail up the St. Johns River to Picolata, where they would unload two infantry regiments and a section of artillery. Those troops would march overland to attack St. Augustine from the rear while Du Pont's gunboats battered the old city from the sea. Before the plan could be set in motion, however, the Federal Navy discovered that the Rebels had abandoned St. Augustine, and Com. C. R. P. Rodgers went ashore "to accept the surrender the city."[26]

Paul Arnau, the mayor of the city, refused to be a party to the capitulation of his beloved town, so Christobal Bravo, acting as mayor pro tem, did the honors. Like most towns of any size in the Deep South, the citizens had

[24] "Movement and Spirit of the War"; "More Arms"; "Slaughter of the Enemy Near St. Mary's, Georgia." See, generally, Hickox, "A Brief History of Clinch's Regiment."

[25] *ORA*, ser. 1, vol. 6, 132–33; Johannes, *Yesterday's Reflections*, 323–24.

[26] *ORA*, ser. 1, vol. 6, 248; East, "St. Augustine during the Civil War," 78–79.

divided opinions regarding Yankee occupation. Cmd. Rodgers, after a few days at St. Augustine, stated, "[M]any [St. Augustine residents]...are strongly attached to the Union, many more are silently attached to it, and most are disposed to quietly acquiesce to the present condition of things." All of the occupation forces agreed that "[t]he women are the most violent secessionists...[and] a party of them cut down the flag pole...the night previous to the arrival of our troops."[27]

The lack of outward resistance led many in the North to conclude that the Rebellion had run its course in Northeast Florida. "Let our affairs prosper in Florida for another month...." a Western journal crowed, "[and] the body of the state...[will return to] Union sympathy and alliance." A journalist in New England went a step further, predicting, "The rebels are beginning to understand the resources and resolution of the [Northern] nation; when these are fully understood among the deluded creatures who have been duped into treason, they will probably make an end of the war and the rebel leaders at the same time." Not to be outdone, a Pennsylvania newspaper declared,

> Florida now takes her position with Delaware, Maryland, Missouri, Tennessee, and Western Virginia, among the loyal slave states, redeemed from the grasp of treason, for not only have we acquired military domination in Florida, but the people are acquiescing in the new condition of things quite readily. They give their adhesion to the Union, and seem to be pleased to be relieved of the despotism of the rebels.[28]

Events already unfolding fifty miles to the south, however, would reveal those optimistic predictions might have been somewhat premature.

New Smyrna

It did not take the Federal Navy long to discover the secret of Mosquito Inlet and New Smyrna. On the morning of 22 March, two US warships, the *Penguin* and *Henry Andrews*, arrived at the entrance to Mosquito Inlet, but the expedition hit a snag almost immediately. The *Henry Andrews* succeeded in

---

[27] East, "St. Augustine during the Civil War," *FHQ*, 78–79; "The Capture of St. Augustine, Florida," *Hartford [CT] Daily Courant*, 22 March 1862, 3.

[28] "Whose Is Florida?" *San Francisco [CA] Bulletin*, 25 March 1862, 2; "The Surrender of Florida," *Danville [VT] North Star*, 12 April 1862, 2; "Another State Redeemed," *Philadelphia North American*, 31 March 1862, 2.

crossing the bar into the lagoon, but the *Penguin*, which drew more water, could not traverse the obstruction. The two commanders, Lt. T. A. Budd and Acting Master Henry Mather, conferred and decided to lead a raiding party to explore the area around New Smyrna. An unidentified Black man, who knew the area well, volunteered to guide the expedition. The officers assembled five launches, taken from the two ships, and a crew of about fifty sailors. Budd led the foray with Mather serving as his second-in-command. All the men carried revolvers or muskets. It appears that a swivel gun (a small bore cannon attached to the front of the launch, which could be aimed by manually shifting it on its swivel) was the Budd party's only artillery (though a couple of sources reported they also took along a 'rifled howitzer.')"[29]

The Union sailors began the operations "in high spirits" and initially believed they could not fail to accomplish their objectives. Their first success came almost immediately. The Federals the blockade-runner *Katie* at anchor in the Inlet and put a picked crew aboard the discovered vessel. The bluecoats then rowed several miles downriver to destroy a salt-making operation south of the entrance to Mosquito Inlet. "Pulling up to Mosquito Lagoon," one raider wrote, "a narrow creek densely wooded on either bank, they found the salt works, about ten miles from the entrance, destroyed them, and started to return."[30]

Thus far the raid had been a lark, and an unnamed Rebel soldier suggested, "It never entered the heads of the [Union] commanding officers that there could be any Confederate force in the vicinity." As a result, the sailors, while returning to their gunships, lost all semblance of military discipline. "Capt. Budd, hoisted his sail and with a fair wind, ran ahead of the other boats, perhaps a mile and a half," a Union raider noted. Concealed nearby, however, in a dense grove of live oak, lay about 150 Rebels under the command of Capt. Daniel Bird. The Confederate troop consisted of Cos. E and H of the 3rd Florida Regiment, led by Bird and Capt. Matthew Strain; thirty horsemen of the Marion Dragoons, directed by Lt. W. E. Chambers, and perhaps twenty-five Clay County bushwhackers from George Huston's band. As soon as Budd stepped ashore, a withering fire erupted from the woods.

---

[29] *ORA*, ser. 1, vol. 6, 111–12; *ORN*, ser. 1, vol. 12, 645–46; "Latest from Port Royal," *New York Tribune*, 7 April 1862, 8; Sweett, *New Smyrna*, n.p.; Cox, "Skirmish at New Smyrna."

[30] "Later from Port Royal"; *ORN*, ser. 1, vol. 12, 645–46; Cox, "Skirmish at New Smyrna."

"The first [Union] boat struck the beach and a man jumped out to pull it up," one eyewitness reported.

> Captain Bird [then] rose up and ordered them to surrender. "Get in," shouted Budd, "Shove off," and turning to Bird, added, "Go to Hell!" The latter [Bird] fired and killed Budd on the spot, and his [Bird's] men poured in a volley that killed or wounded every man in the first boat, except the negro pilot. The boat containing Mather drifted off and lodged on a sand bar. Mather, though wounded, got out to push it off. He was ordered [by the Rebels] to get back but on refusing, was fired on and killed. The second launch fared no better. The remaining launches pulled off for the opposite shore under a hot fire of the Confederates, which killed and wounded several more…[who] found cover in the mangrove bushes there. After dark, the survivors came out and regained their vessel [the *Penguin*].

With admiration for Mather's courage, Bird reported, "[H]e (Mather) fought like a lion; never tried to save himself at all." Meanwhile, the *Henry Andrews* unleashed an artillery barrage into the swamp but did little damage except to the surrounding hardwoods.[31]

Pro-Confederate newspapers listed the death toll as forty Unionists who died during the initial firefight, but the true casualties likely numbered ten dead and several wounded. Many members of the raiding party took cover along the "mangrove-bordered" shore, cowering there until nightfall. Under a shroud of darkness, they escaped by "wading marshes and swimming creeks." Huston's brush-fighters either captured the Negro pilot or took him from the regulars and immediately hanged him. (A modern source asserted that the former slave had originally belonged to Huston and had recently run away from his master.) By daylight most of the Federal survivors had escaped.[32]

That morning the Union warship sent a flag of truce to Bird, requesting permission to recover the bodies of their slain comrades. Bird assented, but

---

[31] Sweett, *New Smyrna*, n.p.; "Later from Port Royal"; Cox, "Skirmish at New Smyrna."

[32] Sweett, *New Smyrna*, n.p.; "From New Smyrna, Fla," *Augusta [GA] Daily Constitutionalist*, 30 March 1862, 1; Fitzgerald, *Volusia County*, 92; Buker, "The Inner Blockade," 75–76.

only if the Yankees returned all the contrabands (enslaved African Americans) they had taken on board. The naval commander agreed to those terms, apparently giving no thought to the fate of the unfortunate Black men. A Federal sailor wrote, "The commanding officer, a Captain Bird...made a show of courtesy by returning [Budd's] papers and watch as if ashamed of the mode of warfare [they had used]." With the swap completed, the Union gunboats quickly departed the area, and the news of the successful ambush on the Federal raiders at New Smyrna briefly lifted the spirits of Florida's Confederates.[33]

JACKSONVILLE

A Federal flotilla comprising both naval and infantry forces left Fernandina on 7 May, heading for Jacksonville. Lt. Thomas Holdup Stevens, a veteran of more than twenty years in the US Navy, headed the task force. Stevens's flotilla included the gunboats *Ottawa*, *Pembina*, *Seneca*, *Huron*, and steamers *Isaac Smith* and *Ellen*. Eight companies of Col. T. J. Whipple's Fourth New Hampshire Regiment would serve as Stevens's infantry contingent. The Yankee expedition hit a snag almost immediately. On 8 May, the vessels arrived at the mouth of the St. Johns River, but due to the "treacherous and shallow water," it took the Union sailors three days of tedious sounding and mapping for the ships to locate a way through the shoals. Rebel scouts, who had been shadowing the Yankee boats since they left Amelia Island, "transmitted the news [of the delay] to Jacksonville over a temporary telegraph wire."[34]

> The wildest disorder engulfed Jacksonville as residents waited for the axe to fall. Hardcore Rebel sympathizers took advantage of this extra time to move to safer pro-Confederate towns such as Lake City, Ocala, or even Tallahassee. Many of Jacksonville's political and business leaders, including Mayor Halsted Hoeg and Otis L. Keene (who managed "Jacksonville's premier hotel"), were natives of the North and welcomed the return of Yankee rule. A group of Confederate adherents loudly proclaimed their intention to kill Hoeg, Calvin Robinson, Judge Phillip Fraser, and several other vocal Unionists. A few of the

---

[33] Sweett, *New Smyrna*, n.p.; "From New Smyrna, Fla"; "The Late Skirmish of the Floridians with the Yankees at New Smyrna," *Charleston [SC] Mercury*, 8 April 1862, 1; Cox, "Skirmish at New Smyrna."

[34] Schafer, *Thunder on the River*, 51–54; "From Fernandina and Jacksonville," *Hartford [CT] Daily Courant*, 31 March 1862, 3.

area's most die-hard Rebels decided that the best solution involved torching the entire town. A recent historian reported, "Jacksonville vigilantes—by then called regulators—were joined by the more bitter and angry of the Fernandina refugees—[and] began clamoring for the destruction of the entire city." Despite the threats and danger, most Jacksonville residents stayed, determined to ride out the coming storm.[35]

After evacuating Amelia Island, the 1st Florida Special Battalion had halted at Baldwin, where Yulee's Florida Railroad intersected the Florida Atlantic & Gulf Central Railroad line (that continued eastward to Jacksonville). While there, Maj. Charles Hopkins received orders from Gen. Trapier to move quickly to Jacksonville and "destroy sawmills, lumber and machine shops" owned by Unionists or Northern-born citizens. Hopkins hastened to comply with the directive, as the Federal boats were finally moving into the St. Johns River and would almost certainly arrive the following day. On 10 March 1862, Hopkins formed his men in a hollow square near the Jacksonville railroad depot and addressed them. He stressed that his troops were "not to molest civilians or damage properties other than those designated." Then, the men moved quickly to raze their targets.[36]

What happened next depends on whom you believe. Northern newspapers generally blamed the subsequent damage on the River City on Rebel regulators and guerrillas, who "were pillaging and destroying all the properties of suspected Unionists…[and] these regulators burned a large foundry, several sawmills, 5,000 feet of lumber, a large hotel [the four-story Judson House], and a dry goods warehouse." Ironically, one of the houses threatened by the flames belonged to Lt. Miles G. Murphy of the 1st Florida Battalion. Hopkins's men, naturally, blamed the conflagrations on "a mob of civilians [who] became aroused by the flames and began burning buildings not on Hopkins' list." Whatever the truth of the matter, the citizens of Jacksonville endured a night of terror and vandalism. Whether Hopkins's unit simply followed orders or engaged in wanton destruction, they certainly added no laurels to their record at Jacksonville. Once the frenzy of destruction started,

---

[35] Schafer, *Thunder on the River*, 50–57; Martin and Schafer, *Jacksonville's Ordeal by Fire*, 71–75.

[36] "From Fernandina and Jacksonville"; Hillhouse, *Heavy Artillery & Light Infantry*, 24–25; Proctor, "Jacksonville during the Civil War," 349; Schafer, *Thunder on the River*, 56–61.

they did nothing to restore order, and as they departed the 1st Florida even fired a volley out the train windows at a group of unruly civilians.[37]

Meanwhile, Rebel zealots J. C. Hemming, and his son Charles took advantage of Stevens's delay to conceal a pair of blockade-runners trapped on the St. Johns by the approach of the Union fleet. The father-son team took the two vessels—the famed yacht *America* (renamed *Memphis* by her new owner) and the steamer *St. Mary's*—almost one hundred miles upriver, scuttling *America* in Dunn's Creek, after removing her masts and sails. They then concealed the *St. Mary's* in nearby Haw Creek. Gov. Milton immediately dispatched Lt. Winston Stephens and his cavalry company to protect *America* until additional Confederate troops could reach the area. The state's chief executive then ordered Capt. John W. Pearson and his Oklawaha Rangers to proceed immediately to the east bank of St. Johns River to prevent the bluecoats from raising the *Memphis*.[38]

The Union ships began arriving at Jacksonville around 10 a.m. on 11 March. The Federals had originally intended to remain there just a few hours. Du Pont and Gen. Horatio G. Wright had agreed prior to the expedition that "the permanent occupation at this time would not be judicious." Lt. Stevens had orders to "land and occupy Jacksonville and other points [along the upper St. Johns] for a few hours for the purposes of reconnaissance or other necessary service." Stevens gave his bluecoats specific orders to destroy any remaining real or personal property that might be of benefit to the Rebel war effort.[39]

White flags flew from many houses along the waterway as the flotilla proceeded up the St. Johns. Shortly after the *Ottawa*, the flotilla's flagship, dropped anchor, Jacksonville sheriff Frederick Leuders came aboard to present the official surrender of the city. Leuders found, however, a delegation of about a dozen citizens already there, pleading with Lt. Stevens to occupy

---

[37] "From Fernandina and Jacksonville"; Hillhouse, *Heavy Artillery & Light Infantry*, 24–25; Schafer, *Thunder on the River*, 56–61.

[38] Schafer, *Thunder on the River*, 56–61; Waters, "Florida's Confederate Guerrillas," 137–39; McCarthy, *St. Johns River Handbook*, 79. Dunn's Lake is today called Crescent Lake. Charles C. Hemming later served in the 3rd Florida Regiment, was captured at Missionary Ridge, escaped Rock Island Prison, became a CSA agent in Canada, and, finally, rejoined his unit in April 1865. A wealthy banker in Colorado postwar, he was instrumental in the installation of the confederate memorial in Jacksonville. See Hemming, "Confederate Odyssey," 69–84.

[39] *ORA*, ser. 1, vol. 6, 239; Schafer, *Thunder on the River*, 52.

Jacksonville permanently. The group stressed that Northeast Florida overwhelmingly held Unionist convictions, and they averred the region's citizens would welcome the bluecoats as saviors. Despite the specific orders from Du Pont, Stevens decided to exercise his discretion and stay in the River City a few days. Stevens stationed members of the Fourth New Hampshire at various locations throughout the town to prevent raids by Confederate soldiers or bushwhackers, who were rumored to be lurking in the nearby woods.[40]

After posting his men, Col. Whipple proceeded to get roaring drunk. He stumbled around town with a "negro woman on his arm" and in his besotted condition "made drunken speeches to his men until daylight." Stevens quickly removed the New Hampshire officer, transporting him to Fernandina under armed guard. Lt. Col. Louis Bell then took command of Stevens's unit, an act that caused considerable grumbling and bitterness among the infantry rank and file. Many of the bluecoats groused that Bell had received his officer's commission only because of his political connections. Unfortunately, Whipple's misconduct seemed mild when compared to what happened next. According to various reports,

> Capt. O'Flynn's Irish company, belonging to the New Hampshire Fourth Regiment, got hold of some whiskey...and...got as drunk as possible. Result: a general row. The officer of the day, Capt. Clough, undertook to quell the disturbance, but he was met with brickbats, clubs, and even bullets. He ordered the guard to fire upon them [the Irish unit], and they did. Some of [O'Flynn's] company returned fire, and some twenty shots were exchanged, resulting in the killing of Mr. Oren Stanton, a member of the company who had been in no way connected with the mutiny. Capt. Greenleaf [leading Co. B] rushed in and charged bayonet...which soon quieted them [the mutineers].... The [Irish] Company were [then] stripped of their arms.

Like Col. Whipple, Stevens had the sons of Eire confined aboard a ship and transported to Fernandina.[41]

The day after the Union troops occupied Jacksonville, Lt. Stevens began searching for the *America* and the *St. Mary's*. Yankee sympathizers had made

---

[40] Proctor, "Jacksonville during the Civil War," 349.

[41] Schafer, *Thunder on the River*, 64–65; "Mutiny in the New Hampshire Fourth," *Boston Traveler*, 7 April 1862, 1; "Col. Louis Bell."

the Federals aware of the Hemming's actions, and the Union sailors conducted two unsuccessful searches of the St. Johns and its tributaries. However, on a third voyage, near Orange Mills, "a small boat was discovered and chased. The three people who were in her pulled ashore and escaped, leaving behind a bag of letters, in which was the very information of which they were in search—the place of *America*'s concealment." Stevens's men quickly located the yacht and began raising her. Capt. Winston Stephens and his company, lurked in the nearby woods, but the officer refused to let his company fire on the Yankees. "I can't shoot them," he said. "I just can't do it—it would be murder." He then led his unit away from Dunn's Creek. As a result, the bluecoats managed not only to refloat the vessel but also to recover the blockade-runner's mast and sails. On their return trip, the Yankees stopped at Palatka and plundered the plantation of William D. Mosely, Florida's first governor. A witness reported that the bluecoats "confiscated" the contents of his smokehouse and everything else that caught their eyes.[42]

Shortly after Lt. Stevens's men took their prize back to Jacksonville, Capt. John W. Pearson arrived on the east bank the St. Johns. Like many of Florida's best guerrilla leaders, he had been born in South Carolina. A resident of Marion County, Pearson had become a wealthy planter, slave owner, and cattleman. He also co-owned, with former senator Yulee, Orange Springs, a famed natural spa and adjoining hotel that had become "popular with Northern visitors." An avid Rebel, Pearson had raised and armed his own company, nicknamed the Oklawaha Rangers. Discovering that *America* had already been recovered, Pearson moved quickly north toward Orange Mills, the home of Dr. R. G. Mays. The Northerners had threatened to kill Mays—describing the physician as a "most malignant Rebel"—in retaliation for the disappearance of a Federal collaborator. Pearson spotted a US gunboat anchored mid-river, near Cole's Mill, and set a trap to capture the enemy and their vessel. The Rebel captain failed to spring his ambush but watched with increasing fury, as White and Black collaborators rowed out to the ship, providing intelligence to the enemy. After the gunboat left, "Pearson arrested four white Unionists and hanged a black slave for providing information to

[42] "Capture of the *America*," *New Lisbon [WI] Juneau County Argus*, 16 April 1862, 3; Waters, "Florida's Confederate Guerrillas," 136–37; Schafer, *Thunder on the River*, 65–66.

the enemy." A modern writer observed, "captured contrabands [Negroes] received no mercy from the enemy [Confederates]."[43]

In his report, both Pearson and Col. F. L. Dancy (who accompanied the guerrilla band) painted a dismal picture of the situation east of the St. Johns. "I regret very much to report to you that at least three-fourths of the people on the Saint Johns River and east of it are aiding and abetting the enemy," Pearson wrote; "we could see them at all times through the day communicating with the [enemy] vessel[s] in their small boats. It is not safe for a small force to be on the east side of the river; there is a great danger of being betrayed to the enemy." Pearson ended his report by suggesting Milton establish martial law in the East Florida counties, describing the area "a nest of traitors and lawless negroes."[44]

On that one point, the Federal commanders would have agreed with the Rebel guerrilla. At first glance, the citizens of Northeast Florida seemed cooperative and eager to return to the Union and the protection of the Federal government. Gen. Thomas W. Sherman arrived at Jacksonville on 19 March, and a delegation of the town's leading citizens "proclaimed their anxiety for the restoration of the United States Government and assured Sherman that a similar sentiment was widespread through this region of the state." Jacksonville civilians, after receiving repeated assurances of continued protection from the US military, sent messages to Fernandina, St. Augustine, and smaller villages in the neighborhood calling for "a convention of all loyal citizens" to form a Union government along the St. Johns River.[45]

In Tallahassee, Gov. Milton pulled out all the stops to make life miserable for the invaders and their allies. Milton convinced Richmond to recall Gen. Trapier, whom pro-Confederate residents blamed for the "night of terror" and the bungled defense of Jacksonville, and selected Brig. Gen. Joseph Finegan to replace Trapier. Though born in Ireland, Finegan had lived in Northeast Florida for much of his adult life and had the respect of people in both Fernandina and Jacksonville. He and Sen. Yulee had been business partners in the Florida Railroad, but he also owned a Nassau County sawmill and

---

[43] *ORA*, ser. 1, vol. 53, 224–26; Waters, "Florida's Confederate Guerrillas," 134–39; Buker, "The Inner Blockade," 75.

[44] *ORA*, ser. 1, vol. 53, 224–26; Waters, "Florida's Confederate Guerrillas," 134–39.

[45] "Unionism in Florida," *New York Times*, 31 March 1862; Schafer, *Thunder on the River*, 68–72.

practiced law in both Fernandina and Jacksonville. Milton immediately dispatched all the troops he could muster to the region—according to the Federals, at least three thousand men.[46]

Col. W. S. Dilworth, commanding the 3rd Florida Regiment and various smaller units around Jacksonville, made a quick reconnaissance and decided assaulting the Union gunboats at Jacksonville would be nothing short of suicide. Instead, he ordered his troops to adopt guerrilla tactics, or, as Dilworth called it, "a system of annoyances." Vessels going up and down the St. Johns almost immediately came under heavy fire from the shore. Capt. Winston Stephens's company and other scattered irregular warriors began sniping at the Union sailors. A Federal mariner described a typical attack: "Now and then a volley from a thicket was poured into a boat crew…[on] some narrow and winding waterway, and men fell mortally wounded, but this was the work of scattered guerrillas and not organized troops." While the tactic did not halt Yankee voyages along the St. Johns and its tributaries, it proved a significant annoyance to Pres. Lincoln's brown water navy.[47]

Almost immediately, the land war heated up around the River City. Dr. Samuel Proctor wrote, "On the evening of March 24, Confederates captured two Federal pickets who strayed beyond their defense lines. Early the next morning, Confederate Lieutenant Thomas E. Strange—supported by ten volunteers—attacked Federal pickets at Brick Church along the western edge of town. Three Federals were killed and four were captured." A member of the Fourth New Hampshire described his unit's losses on a single day.

> On the 24th…company H, on picket, were surprised by 60 guerrillas, and Geo. W. Goldsmith was instantly killed and Richard E. Davis dangerously wounded. Corporal Austin Wallace, Wordsworth, and Levi Martin were taken prisoner. James Fletcher and Andre Collins escaped…Charles McQuesten and Solomon Bumford were [also] taken prisoner on the 24th, while in the woods for water.[48]

---

[46] Schafer, *Thunder on the River*, 71–72; Regarding Gen. Finegan, see generally, Litrico, "Joseph Finegan," 23–27.

[47] Schafer, *Thunder on the River*, 80; "Another Fight with Gunboats," *Augusta [GA] Daily Constitutionalist*, 1 June 1862, 1, contains an account of the sniping by Stephens's men.

[48] "The Civil War Fifty Years Ago," *Lexington Leader*, 9 March 1912, 9.

Two Union expeditions designed to drive off the partisans proved dismal failures, but Dilworth's "system of annoyances" made Yankee pickets nervous, which led to an additional tragedy. Anxious outposts noticed a strange group lurking in the woods near Federal lines and fired into them. The hidden party proved to be a large group of runaway slaves, who had escaped from their masters near Lake City and traveled through fifty miles of "enemy territory," looking for a chance at freedom.[49]

Instead of abiding by their promise to stay and protect the region, on 9 April, the Union military boarded their ships and left Jacksonville. One of their last acts was the burning of Orange Mills. The bluecoats agreed to take a number of outspoken Unionists with them, and many Northern-born loyalists took that opportunity to return to their home states. A sizeable minority of ordinary citizens escaped from Jacksonville, with its bushwhackers and regulators, by taking up residence in vacant houses at Fernandina, where they subsisted on handouts by the Union military.[50]

Members of the US Congress, infuriated by the abandonment of Jacksonville, requested Pres. Lincoln to provide "all facts and circumstances...in regard to the late evacuation of Jacksonville." Lincoln cryptically noted that the Union troops left the city under orders from the commanding general "for reasons which it...is not deemed compatible with the public interest at this time to disclose." Officially, that ended the matter. However, Jacksonville resident Calvin Robinson reportedly discovered later that Lincoln considered leaving the town a tremendous blunder. However, the president publicly stood by his generals and Adm. Du Pont, whom he could not afford to alienate.[51]

An even greater betrayal occurred at Jacksonville that involved not White Unionists but the hundreds of slaves who had risked their lives to flee their masters for an opportunity of gaining their freedom. In defense of the Federal officers at Jacksonville, the US government had issued no directives regarding the handling and disposition of the "contrabands." Earlier in the

---

[49] Proctor, "Jacksonville during the Civil War," 350–51; "N.H. Jacksonville: March," *Amherst [NH] Farmer's Cabinet*, 10 April 1862; Schafer, *Thunder on the River*, 79–82.

[50] Schafer, *Thunder on the River*, 74–76; "Evacuation of Jacksonville by Union Troops," *New York Tribune*, 21 April 1862, 3.

[51] Schafer, *Thunder on the River*, 75–76; Robinson, "Account of Some of My Experiences."

war, a nearly unanimous Congress had passed a resolution "affirming that the purpose of the War was to preserve the Union and not to change Southern domestic institutions [slavery]." As a result, the Yankee officers followed the dictates of the much-reviled Fugitive Slave Act and returned the Black runaways to their masters before departing Jacksonville. Union officer Rufus B. Saxton, an outspoken abolitionist, described an incident he witnessed just before the boats set sail. Col. Henry Theodore Titus, who had fought with the proslavery forces in Bleeding Kansas, arrived in Jacksonville to reclaim one of his slaves. Saxton wrote, "On receiving the negro[,] Col. Titus put a slipping noose rope around the negro's neck, with a timber pitch at his arm, mounted his horse and dragged the poor victim off in the presence of our troops." Before leaving Jacksonville, Union officers also restored all runaway slaves to their masters, including fifty-two who had "reached Fernandina." Union policies regarding slavery would change dramatically in the coming months, but it came too late to help the runaways in Northeast Florida during spring 1862.[52]

However, before leaving Duval County, "Du Pont left some gunboats at Mayport. He reasoned that it took fewer ships to blockade the river when stationed at its mouth than if his ships were on station at sea.... [W]inter storm would endanger, and occasionally drive off, his ships at sea, weakening the blockade; ships at the mouth of the river could maintain a tighter blockade under safer conditions." If they had remained at Mayport, most of the St. Johns would have remained under Rebel control, but "the true 'Inner Blockade of Florida' began when the naval vessels steamed up and down the St. Johns disrupting Confederate activities all along the river."[53]

St. Augustine

The Unionists had taken St. Augustine without firing a shot, but holding the Ancient City proved a more formidable task. Stephen V. Ash, a modern historian, has theorized that "three unique worlds existed in the occupied South: garrisoned town, no-man's land, and the Confederate frontier [an area primarily dominated by the Rebel military or guerrilla bands]." St. Augustine fit that pattern exactly. The municipality, with the cannons at Fort Marion and Federal gunboats made the city practically invulnerable to any force Florida

[52] De la Cova, *Colonel Henry Theodore Titus*, 178–79. The city of Titusville took its name from Col. Titus.

[53] Buker, "The Inner Blockade," 77.

could muster against it; but any Federal soldier took his life in his hands to venture very far outside the city walls.[54]

Within two months of its capture, Confederate irregulars turned the outskirts of St. Augustine into Rebel territory. "Several Unionists were hung in St. Johns County," East reports, "where guerrillas even grew bold enough to slip into St. Augustine to snipe at Federal sentinels," Despite the repeated claims by Union politicians and military leaders that most East Floridians were loyalists at heart, Col. Louis Bell discovered "a constant [line of] communications was kept up between the inhabitants of this city and the enemy." The Federals tried to fight back as best they could. The Federals burned the Fairbanks house, within five miles of St. Augustine, because they spotted guerrilla horses there on several occasions. Despite such harsh actions, it did nothing to lessen the partisans' "reign of terror."[55]

The occupation force made several attempts to broaden its scope of control but met with little success. For example, in an effort to increase the mobility and effective range of their patrols, Col. Bell decided to turn one of his companies (apparently Co. H, under the command of Lt. Henry M. Hicks) into a cavalry unit. After scrounging some horses and practicing various cavalry exercises, Hicks decided to take his "horse soldiers" into the field. They had not gone very far when a Rebel sharpshooter shot Hicks's animal from under him. Hicks beat a hasty retreat to St. Augustine, much to the amused delight of his men. This event apparently halted Col. Bell's experiments with cavalry. In a second attempt to break the bushwhackers' hold on the countryside, Adm. Du Pont and Gen. H. W. Benham decreed that outrages against Unionists "shall be visited fourfold upon the inhabitants of disloyal or doubtful character nearest the scene of any such wrongs when the actual and known perpetrators cannot be discovered. This seemingly ruthless proclamation could not be carried out...[because] these isolated garrisons could not operate far afield, due to the strength of the...Confederate forces." In the end, the Yankees' draconian measures at St. Augustine, designed to punish the Rebel irregulars, came to naught.[56]

---

[54] Ash, *When the Yankees Came*, 77–92; John Milton to [?], 25 April 1862, Milton Letterbook; see generally, Totten, "Ancient City Occupied."

[55] East, "St. Augustine during the Civil War," 85; *ORN*, ser. 1, vol. 14, 333–34; Totten, "Ancient City Occupied," 57–58.

[56] Totten, "Ancient City Occupied," 43–44; East, "St. Augustine during the Civil War," 85.

The military authorities might have been frustrated by their inability to halt bushwhacker depredations, but the citizens suffered even more from the occupation. An unidentified woman from St. Augustine, writing to a friend in California, explained the situation in Northeast Florida.

> [I]t is not considered safe for residents here to travel the roads or even to venture a few miles out of town. We have no beef in the market, for the butchers are afraid to go out in the country to purchase or drive in cattle; and we are very much troubled to get wood, for none of the people will venture more than three or four miles to cut it. There are forming throughout the country, numbers of guerrillas or marauding parties, after the style of the old Regulators, infesting the roads and waylaying persons who have shown any Union proclivities; driving them off from their places [homes] and threatening to set fire to their building[s].... You may judge from all this what a state the country is in—just about as [bad as it was] in Indian times [during the Second Seminole War]. The [New York] *Herald* of...[April] 20th says: "The whole of Florida is restored to the Union. The capture of Fernandina and St. Augustine, with their defenses, brings back Florida under the folds of the 'Stars and Stripes.'" That sounds very well for a newspaper paragraph; but the fact is they [the Federals] just hold two places, and are not in possession of anything one mile beyond their pickets.[57]

TALLAHASSEE AND RICHMOND (GOVERNMENT ACTIONS)

With hope for victory and enthusiasm for the war already lagging in several of the seceded states, the Confederate Congress—at the urging of Pres. Jefferson Davis—enacted a Conscription Law. The original Rebel army had been an all-volunteer force of men serving one-year enlistments. With their enrollment period ending, the Confederacy faced the real threat of its armies folding their tents and heading home, so the Southern Congress passed a law extending the current terms of enlistments to three years and making all White males between the ages of eighteen and thirty-five subject to the military draft.[58]

---

[57] "The Condition of Florida," *San Francisco [CA] Bulletin*, 26 June 1862, 2.

[58] Martin, "Civil War Conscription Laws"; Reiger, "Deprivation, Disaffection, and Desertion in Confederate Florida," 286.

The exemptions from service, written into the new law, made it extremely unpopular with many Southerners. One of the clauses that caused the most resentment allowed a person with enough money to hire a substitute and thus avoid military service. A second article exempted owners of twenty-five or more slaves and their overseers from enlisting if they so desired. Cries of "rich man's war, but a poor man's fight" soon resounded throughout the Confederacy. Of course, food to supply the army had to be grown, and while many in the Virginia and Tennessee armies went hungry, exempted planters raised hundreds of acres of inedible, but valuable, cotton. Despite their disgust at the inequities, most enlisted men obeyed the law and continued to serve their country.[59]

In his first few months as Florida's governor, John Milton reached two important conclusions; first, that Jefferson Davis's government would syphon off every man (of military age), cow, and sack of salt in Florida, and give the state nothing in return. Despite those misgivings, Florida never failed to meet its quota of men for the army. His second conclusion was that from mid-1863 till the end of the war, cattle and salt, provided almost exclusively by Floridians, would be the primary source of food for the Rebel armies in the eastern theaters.[60]

More importantly, Milton also determined that guerrilla warfare offered his state's best hope of keeping the interior of Florida in Rebel hands. He wrote: "[O]ne thousand men divided into small companies—well-armed—acting as Guerrillas or Rangers and ably commanded—can do more to defend Florida from the enemy than thousands in regular service." Milton began quietly organizing a cavalry company for that purpose. William A. Owens, a South Carolina native and "one of the largest planters in the state," began by raising a unit from Marion County called the Marion Dragoons. (Marion County, with its large number of South Carolina natives, remained one of the most "Confederate" counties in the state throughout the war and into the twentieth century.) The dragoons would form the nucleus for a troop later designated as the 2nd Florida Cavalry (CSA) Regiment. Due to age and infirmity, Owens soon retired. Col. Carraway Smith assumed leadership of the unit, with A. H. McCormick as his second in command. After Owens's departure, the dragoons (who numbered almost two hundred) were divided into

---

[59] Martin, "Civil War Conscription Laws"; Reiger, "Deprivation, Disaffection, and Desertion," 67–71.

[60] Wynne and Taylor, *Florida in the Civil War*, 75–78.

two companies. Capt. Samuel Rou directed the first unit, and Capt. William E. Chambers directed the other. Gov. Milton recalled Chambers's partisans, initially stationed at Fernandina, to the mainland just before the Union gunboats landed on Amelia Island.[61]

Early in 1862, the Confederate Congress also enacted a law that allowed states to organize groups of "Partizan Rangers." According to the decree, the men in the units had to be at least thirty-five years old. The law allowed these soldiers to turn in weapons and ammunition captured from the enemy for cash, making the rangers a type of "land-based privateer[s]." Maj. T. W. Brevard, a Tallahassee attorney and politician, organized the 1st Partzan Ranger Battalion to serve in East Florida. The companies hailed from Duval (two), Putnam, Bradford, Hamilton, Jefferson, Jackson, and Washington counties. Milton also formed a few greybeard squads—as a stopgap—until more reliable troops could be sworn into service. "I organized," he wrote, "the old men [in Hernando County]—under J[ames] Baker for the purpose of keeping the negro population under control." He also authorized Capt. Thomas B. Law to form a similar group at Lake City. (The fact that a large group of African Americans had reached Jacksonville from Columbia County indicate the ineffectiveness of thosc units.)[62]

Crystal River

Early in the war, a number of Northern newspapers published articles describing various important Southern ports, but none of these dispatches mentioned Crystal River. That oversight was completely understandable. After all, the waterway measured only thirteen miles from its headwaters to its mouth, had a shallow entrance, and meandered through a marshy, sparsely settled region. Despite its miniscule length and remote location, the region witnessed several engagements during the war. Joe Knetsch noted,

> The entire coast of [modern] Citrus County was a hive of activity. The island-studded coastline was perfect for eluding the larger

---

[61] John Milton to [?], 23 April 1862, Milton Letterbook; Evans, ed. (and J. J. Dickison), *Confederate Military History* (Florida), 11:48–50.

[62] John Milton to [?], 13 March 1862, Milton Letterbook; John Milton to [?], 3 April 1862, Milton Letterbook; "Partisan Rangers," *Charleston [SC] Courier*, 30 May 1862; Waters and Edmonds, *Small but Spartan Band*, 122–23; Hillhouse, *Heavy Artillery & Light Infantry*, 94–97. Almost without exception, Floridians ignored the age limits in the Partizan Ranger Law.

> ships of the [US] blockading squadron. Its many rivers, run conveniently inland about ten to fifteen miles, [and are] connected by roads improved by the [US] Army during the Second Seminole War (1835–1842), made the county a haven for blockade running.[63]

In January 1862, the *Kate* had escaped the Federal capture of Cedar Keys by taking refuge at Crystal River, but the first recorded action there occurred on 24 April 1862. A few weeks earlier, Capt. Paul Ardisson, commanding the blockade-runner *Lahorn Eagle*, left Havana, Cuba, bound for New Orleans with a "large quantity of [gun]powder." Off the Louisiana coast the USS *Tahoma*, commanded by Flag Officer William M. McKean, spotted Ardisson's vessel and pursued it. The chase lasted several days, and as the *Tahoma* closed in on the *Lahorn Eagle*, Ardisson and his crew abandoned ship and rowed for the mouth of Crystal River. Before leaving his ship, Capt. Ardisson fired his blockade-runner, and as the crew from the *Tahoma* neared her prize, the *Lahorn Eagle* exploded. McKean stated in his report that "she was literally blown to atoms." Ardisson and his crew escaped up Crystal River, and the captain soon returned to the lucrative blockade-running trade.[64]

A second encounter, only a couple of weeks later, took a more deadly turn. Chasing another fleet ship, the *Tahoma* called for assistance, and the USS *Beauregard* quickly joined McKean's vessel on patrol. When the unidentified blockade-runner took refuge in Crystal River, Capt. David Stearns, of the *Beauregard* took several men in a launch, and rowed into the waterway. Capt. Sam Hope, a prewar surveyor, waited in ambush with five members of his "Coast Guard" unit and fifteen volunteers from the area. The Rebels' first volley killed two men, including Capt. Stearns, and wounded a couple of more. The survivors were captured and sent to prison in Tallahassee. Some of Hope's volunteers wasted no time in hanging a captured Black man who had guided the Federals up Crystal River. An unidentified Floridian wrote, "Hanging would be the penalty inflicted upon a white man [for] deserting to

---

[63] "Crystal River National Wildlife Refuge"; Knetsch, "Armed Boats off Crystal River," 6–7. During 1861–1865, modern Citrus County was part of Hernando County.

[64] Knetsch, "Armed Boats off Crystal River," 6–7; *ORN*, ser. 1, vol. 17, 290–92; "Affair at Crystal River," *Charleston [SC] Courier*, 16 July 1862, 4.

the enemy, and we see no good reason why a negro should not suffer in the same manner for the same offence."[65]

Meanwhile, the *Beauregard* had chased "a sloop-rigged vessel" that sailed out of Crystal River shortly after Stearns's launch entered the stream. The Union ship, however, lost the blockade-runner in a sudden squall. Master's mate William H. Melson, of the *Beauregard*, reported, "Next morning...[we] sent an armed boat [into Crystal River] and made a thorough search, firing muskets at intervals for six hours but without success. [On the] same night [we] got underway and shifted positions."[66]

Sen. Yulee sent a message to Gen. Finegan at Tallahassee applauding Capt. Hope for his aggressive defense of the Gulf Coast. The citizens of Florida's capital city also celebrated Capt. Hope's small victory and the arrival of the prisoners. Hope, however, had already shifted his troops south to Bayport, which seems to have been his primary duty station during the early stages of the war.[67]

TAMPA AND SOUTH FLORIDA

The reminders that Tampa remained the last major port in South Florida still in Rebel hands were printed with annoying frequency by Northern newspapers and served as a constant irritant to the bluecoats at Key West. To make matters worse for the Federals, Maj. W. L. L. Bowen, commanding Confederate forces in that region, had proven to be an aggressive and pugnacious fighter. Bowen had often taken the fight to the enemy—capturing numerous fishing vessels, frustrating the blockade, and even raiding Union-occupied Egmont Key. When Maj. Robert Benham Thomas replaced Bowen at Tampa on 10 February 1862, the Unionists apparently decided to test his mettle. He might, they hoped, be pressured or bluffed into evacuating the South Florida city. A Kentuckian and 1852 West Point graduate, Thomas had been stationed at Fort Brooke during the Third Seminole War. There, he met a daughter of wealthy merchant James McKay Sr., fell in love, and got married. Thomas soon resigned his army commission and went to work for his father-

---

[65] Knetsch, "Armed Boats off Crystal River," 6; "Affair at Crystal River"; *ORN*, ser. 1, vol. 17, 290–91.

[66] Knetsch, "Armed Boats off Crystal River," 6; *ORN*, ser. 1, vol. 17, 290–91; "Affair at Crystal River."

[67] Knetsch, "Armed Boats off Crystal River," 6–7; Proctor, *FL100*, 3 July 1862.

in-law. When Bowen departed for the Army of Tennessee, Gen. Finegan appointed Thomas—the hometown hero—to defend Tampa.[68]

The Federals immediately increased pressure on the Confederates in the Hillsboro County area. Shortly after assuming command, Thomas received a petition from citizens of the Clearwater Harbor—Old Tampa Bay region—requesting protection from the increased harassment by Union raiders. With his limited manpower and resources, it proved a difficult plea to honor. In February, a pair of ships from the Blockading Squadron flying Confederate flags, captured the blockade-runners *Spitfire*, *Atlanta*, and *Caroline*. J. E. Whitehurst and a Mr. Girard, local Unionists, urged the Yankees to increase the pressure on Tampa, claiming that forty of their neighbors (within six miles of Fort Brooke) longed for an opportunity to rise up and throw off the Rebels' shackles. Whitehurst and Girard also informed the Union commander at Key West, that the graycoats at Fort Brooke were a ragged, ill-disciplined, and cowardly group who would offer little resistance in case of attack.[69]

The first operation of any consequence, however, occurred on the sparsely populated tip of the Pinellas Peninsula region of Tampa Bay. Before daylight on 12 February, Union warships appeared offshore from Big Bayou, on the southern end of the peninsula. Only two families lived in that rural paradise—John A. Bethell and Abel Miranda. Miranda, an Indian War veteran, fisherman, and cattleman, described his home in the wilderness as "full of game and varmits. That year (1857) I killed eleven bears and one panther; [and of]...ducks, wild geese, and turkeys; I could get all I could make use of and never go half a mile from home." Contemporary sources indicated that he knew the South Florida Gulf Coast waters better than any man alive. Though pro-Confederate in sympathy, he might have been content to sit out the war, making money from whichever side could pay him. All that changed on the morning of 12 February.[70]

Miranda awoke that morning and spotted a couple of gunboats sitting off Pinellas Point. He and his wife immediately surmised that the Yankee

---

[68] "Affairs on the Florida Coast," *Albany [NY] Evening Journal*, 29 April 1862, 2; Waters, "Tampa's Forgotten Defenders," 9–10; Krick, *Lee's Colonels*, 370.

[69] Waters, "Tampa's Forgotten Defenders," 10–12; "The Enemy on the Florida Coast," *New Orleans Daily True Delta*, 20 February 1862, 1; *ORN*, ser. 1, vol. 17, 85; "Our Key West Correspondence," *New York Herald*, 24 March 1862, 5.

[70] Pizzo, *Tampa Town*, 66–67; USF, "100 Years of Black St. Petersburg."

vessel planned to destroy their homestead. The Mirandas gathered their children and ran to the Bethell house. Many years later Bethell wrote an account of the raid. "About 7 a.m., they [the Yankees] opened fire with round shot. They made three good line shots for the [Miranda] house.... That was the first time we heard cannon shot whistle over our heads.... It seemed like the heavens had fallen through and scared us so [badly] that we did not know whether we were killed or paralyzed." After the Federal vessels departed, Miranda and Bethell surveyed the damage. The Yankees' gunfire had completely demolished Miranda's house. He also found that his boats had either been confiscated or destroyed, his citrus grove cut down, and most of the livestock stolen. The farm animals that remained, primarily chicks and piglets, had been maimed, and the two men had to destroy them to alleviate the creatures' suffering.[71]

The reason for the unprovoked attack on the Miranda homestead has never been adequately explained. The reason most often given—that Miranda had lured several Union sailors to their deaths by pretending to be runaway slaves seeking refuge—happened months *after* the attack on the Miranda house. Miranda's niece guessed "that the Federals destroyed...uncle's house out of pure malice." Miranda seemed to have placed blame for the raid on the Unionist Whitehurst family. Whatever the reason, Abel Miranda became a constant terror to the bluecoats. "Miranda was tagged 'The Cat' by the Yankees. He was a tough, powerful man...[who] headed a small independent detachment of coast guards who played havoc on Federal foraging parties." Though not found on any Confederate rosters, a partial list of Miranda's independent unit included Pancho Sanchez, John Bethell, James Barnett, Timothy Buckley, Anderson Woods, John Allen, and James Williams.[72]

The Unionists seemingly increased their depredations in the bay area after the raid at the Miranda homestead. On 29 February, a hundred bluecoats again raided the Shaw Point area (on the southern bank of the Manatee River), questioning and abusing the three or four families living there in an attempt to locate some "big guns" supposedly hidden nearby. As if that were not enough, rumors circulated that the Federals had incited the Seminoles to attack the Rebels at Tampa. Acting quickly, Maj. Thomas "dispatched

[71] Pizzo, *Tampa Town*, 66–67; USF, "100 Years of Black St. Petersburg"; Bethell, "Pinellas: A Brief History"; see generally, W. Michaels, *Making of St. Petersburg.*

[72] Pizzo, *Tampa Town*, 66–67; Bethell, "Pinellas: A Brief History."

George Lewis...to determine the Indians' intentions and placate them if possible."[73]

On 13 April 1862, two Union gunboats, the *Ethan Allen* and *Beauregard*, steamed up the bay, and Lt. William B. Eaton sent Maj. Thomas a formal demand to surrender Ft. Brooke and Tampa. Eaton concluded his message by promising to allow Thomas twenty-four hours to remove all noncombatants. Thomas politely rejected the request to capitulate but thanked the Union officer for his humanity in allowing time to evacuate the women, children, and elderly. The following day the two Federal vessels bombarded the fort in a furious but basically ineffective attack. Neither the fort nor the town sustained much damage, and neither officer would remain long at Tampa Bay. Flag officer W. W. McKean, directing the US East Gulf Blockading Squadron, removed Eaton from command of the *Ethan Allen* for unknown reasons. Thomas, however, received what amounted to a promotion. Gen. Finegan reassigned the former West Pointer to Tallahassee where he served as Finegan's adjutant.[74]

Shortly thereafter, Capt. John W. Pearson, the aggressive guerrilla who had arrived too late to prevent the salvage of Yacht *America*, replaced Maj. Thomas as the Rebel commander at Tampa. The Federals wasted no time in finding out if Pearson had courage and a fighting spirit. The Rebel commander reported, "On Monday morning, June 30 [1862], the gunboat [*Sagamore*] hove into sight...[and] turned his broadside to us and opened her ports, then started a launch with a lieutenant and 20 men, bearing a flag of truce toward the shore." Pearson met the enemy in a boat before they could set foot on his domain and rejected the Unionists' demand to capitulate. Pearson reportedly told the Union officer he had "no understanding of the word surrender." The gunboat began shelling about 6 p.m., and the bombardment lasted for an hour. The graycoats at Fort Brooke responded with counter fire, but most of their shots fell short of the enemy vessel. Several civilians acted as if the fireworks display had been put on for their amusement. The next day the *Sagamore* returned, stayed out of range of the fort's guns, and again shelled the town. A Tampa citizen, writing to a religious journal, thanked "a gracious Providence" that the cannon fire resulted in no

---

[73] Waters, "Tampa's Forgotten Defenders," 10–12; *ORN*, ser. 1, vol. 17, 85; "The Enemy on the Florida Coast"; "Our Key West Correspondence."

[74] Waters, "Tampa's Forgotten Defenders," 12–13; *ORN*, ser. 1, vol. 17, 215–16.

injury "to man, beast, house, or fence." In the wake of the bombardment, Pearson realized the inadequacy of the fort's artillery and sent a couple of his men to his machine shop at Orange Springs to manufacture two rifled cannons. They soon returned to Tampa with the homemade—but effective—artillery pieces, which Pearson named "Tiger" and "Hornet."[75]

CLAY COUNTY

After abandoning Jacksonville on 9 April 1862, Federal gunboats continued patrolling the St. Johns River. It served as a way of "showing the flag," providing encouragement for Northeast Florida Unionists, and collecting valuable supplies, such as lumber, cattle, and cotton. Using the waterway, however, remained a dangerous, contested route for the US sailors and marines, as the guerrillas persisted their practice of sniping at the ships from the shoreline. On 18 May, for example, two Union vessels steamed upriver, heading for the Town of Welaka. James W. Bryant, a vocal Unionist and one of the founders of the village, had obtained passage on one of the ships. (His children did not share their father's political allegiance as he had two sons in the Confederate army, and daughter Olivia was married to Rebel guerrilla leader Winston Stephens.) As the gunboat neared Horse Landing, where the river narrowed significantly, Stephens and his troops opened fire on the vessels, killing twelve men. The bluecoats "beat a hasty retreat," going downriver to Orange Mills where the US sailors "obtained lumber, and after making coffins, they buried their dead at Madison Point." According to one report, thereafter "the gunboats keep shy of Jacksonville...only visiting under a flag of truce when they wish to communicate."[76]

Despite such incidents, Union officers, with the unshakeable faith of zealots, remained convinced that the vast majority of Floridians despised the Confederacy and longed to return to their "first love." The fact that the state had, in the last four months, mustered two partisan ranger battalions, a regiment of cavalry, and numerous as yet unassigned companies did little to alter their opinion. Lt. Daniel Ammens, commanding the USS *Seneca*, wrote,

---

[75] *ORA*, ser. 1, vol. 14, 111; Waters, "Tampa's Forgotten Defenders," 12–13; W. I. Murphy, "Mr. Editor," *Augusta [GA] Southern Christian Advocate*, 14 August 1862; "The Oklawaha Rangers," unpublished mss. [by Pearson's daughter], Zack C. Waters collection, Rome, GA.

[76] Schafer, *Thunder on the River*, 80–83; "Florida News," *Macon [GA] Telegraph*, 30 May 1862, 3.

"[M]ost of the people living along the [St. Johns] river would gladly acknowledge the authority of the United States were they not in fear of violence from bands of Confederate Regulators." This became the Yankees' "party line," and they chose to make an example of George Huston, the violent, diehard Rebel bushwhacker. Adm. Du Pont gave the rationale for the expedition. "I wished him [Huston] a prisoner, also for reclaiming the general tranquility among the persons along the river, most of whom, I doubt not, would gladly acknowledge the Government of the United States, were they not in fear of violence from men of this character."[77]

Huston lived at Claremont Plantation (in Clay County), on the west bank of the St. Johns River, just south of Fleming Island. The Federals chose to make an example of Huston for several reasons. First, he had a "violent hatred for the Yankees, and a fanatical devotion to the South." Secondly, he had participated in the ambush and killing of Budd and Mather, at Mosquito Inlet. This had "left a bitterness with Federal [Naval] crews, and Huston [also] proudly boasted that he had strung up the African American pilot who had led the incursion in Volusia County." Lt. Ammens devised the scheme to kill the desperado, in hopes that it might put the fear of reprisal in the minds of other Rebels.[78]

Lt. J. G. Sproston, described as "an able, brave, and devoted officer from the State of Maryland," led the Union task force. Arriving at Claremont Plantation about dawn, Sproston approached the guerrilla's residence and shouted a demand for his surrender. Outnumbered seventy to one, Huston proved game. "He appeared in his door, armed with a double-barreled [shot]gun, two pistols, and a Bowie knife." The bushwhacker immediately opened fire on his enemies, killing Sproston. A heavy volley followed a second later, and Huston crumpled in a hail of bullets. Seeing her spouse shot to pieces, Huston's wife threw herself upon the body of her husband, which halted the shooting. A weak volley of shots came from the distant woods as Huston's band apparently tried to avenge their leader's death. Though wounded in four

---

[77] Schafer, *Thunder on the River*, 80–83; "Florida News"; "From Commodore DuPont," *Harrisburg [PA] Patriot*, 19 June 1862, 8.

[78] Schafer, *Thunder on the River*, 80–83.

places, Huston survived for four agonizing months before dying in a Union prison.[79]

TALLAHASSEE—GUERRILLA REGIMENT

By mid-1862, the 2nd Florida Cavalry, Gov. Milton's pet project, was nearing completion. The companies, commanders, counties of origin, and nicknames were as follows: Company A, Capt. Clinton Thigpen, Gadsden County, "Milton Rangers"; Company B, Capt. Winston Stephens [replacing Benjamin Hopkins], Putnam and Volusia counties, "St. Johns Rangers"; Company C, Capt. William E. Chambers [replacing W. A. Owens], Marion County, "Marion Dragoons"; Company D, Capt. P. B. Brokaw, Leon County, "Tallahassee Guards"; Company E, Capt. H. T. Blocker, Leon County and South Georgia, [no known nickname]; Company F, Capt. Samuel F. Rou [also replacing W. A. Owens], Marion and Alachua Counties, "Confederate Rangers"; Company G, Capt. W. H. Milton, various North Florida and South Georgia counties, "Harris Troop"; Company I, Capt. John Caraway Smith, various East Florida Counties, "Aucilla Troop"; Company K, Robert H. Harrison, Nassau County, [no known nickname].[80]

The units would remain a conglomeration of independent companies, stationed wherever needed, until 4 December 1862, when they would be consolidated into a regiment. Arming the various groups took time, but as they received weapons, Milton and Finegan immediately put them to work. For example, Capt. Chambers, who had the weapons taken during the fight at Mosquito Inlet, would be dispatched to the Etoniah Scrub, on the border of Clay and Putnam Counties, to investigate rumors of a possible slave revolt. Meanwhile, Winston Stephens and his men continued sniping at Union gunboats on the St. Johns.[81]

---

[79] Schafer, *Thunder on the River*, 81; *ORN*, ser. 1, vol. 13, 83–84; "Naval Officer Killed by Rebel Desperado," *Washington [DC] Evening Star*, 17 June 1862, 3; Buker, "The Inner Blockade," 75–76.

[80] Hartman and Coles, *BRF*, 4:1408–557; Sifakis, *Compendium of the Confederate Armies*, 10–11. See generally, Stoutamire, "Florida's Army," 27–44. The presence of soldiers from South Georgia and South Alabama in Florida units would become more common as the war progressed. Numerous companies in the 10th and 11th Florida Infantry Regiments had soldiers from two or three states.

[81] John Milton to [?], 24 March 1862, Milton Letterbook; Schafer, *Thunder on the River*, 80.

On 2 July, a new leader appeared on the scene who would become Florida's most famous (and arguably the state's most effective) partisan leader. John J. Dickison, a forty-year-old transplant from South Carolina and wealthy Marion County planter, received permission to raise the tenth and final company of the 2nd Florida Cavalry. Dickison had already served as a lieutenant with the Marion Light Artillery, but when he failed to win re-election by that unit, he returned to Florida and discovered his true calling. A recent historian related,

> His company, along with various other units temporarily under his command, patrolled areas along the St. Johns [River] and eastward to the coast, ambushed Union foraging expeditions, and captured pickets and isolated bodies of troops. Even when he was not actually attacking the enemy, fear of Dickison kept the Union forces for the most part bottled up in St. Augustine and a few scattered posts along the St. Johns River. The Federals could occupy the towns, but they were never able to effectively to control the countryside. Nicknamed the "Swamp Fox," "War Eagle," or simply "Dixie," much of the credit for holding on to [the interior of] Florida belonged to one man—John Jackson Dickison.[82]

[82] Weinert, "Swamp Fox of Florida," 4–11. See generally, Koblas, *Swamp Fox*; M. E. Dickison, *Dickison and His Men.*

Chapter 3

"Florida Is Virtually in Our Possession"

(August–December 1862)

THE SITUATION IN MID-1862

The Confederates' military outlook in Virginia improved in the last few months of 1862. Gen. Robert E. Lee's Army of Northern Virginia had driven the Federal armies from the gates of Richmond in the Seven Days campaign and won the battles of Second Manassas and Fredericksburg. Those victories, however, had often been bloodbaths that depleted the ranks of the Rebel Army's earliest, most enthusiastic, and most determined volunteers. The 2nd Florida Infantry, for example, had fought at Williamsburg and during the Seven Days battles, leaving little more than one hundred men available for the next fight. Gov. Milton promptly ordered the 5th and 8th Florida regiments to Virginia's "killing fields." There, the new units joined the remnants of the 2nd Florida in Brig. Gen. E. A. Perry's Florida Brigade.[1]

Richmond's demands for Florida troops for the Army of Tennessee proved even more insistent. The 1st Florida Regiment had fought at Pensacola and Shiloh, but with their enlistments expiring, many of the unit's survivors opted to return home rather than remain with their original outfit. Gov. Milton sent the remaining members of the 1st Florida, along with the 3rd, 4th, 6th, and 7th Florida Infantry, and the 1st Florida Cavalry (dismounted) to Gen. Braxton Bragg's Western Army. Initially posted to East Tennessee, they spent several difficult months guarding the East Tennessee and Virginia Railroad, which ran from Richmond, Virginia, to Chattanooga, Tennessee. Union sympathizers in that region engaged in a bridge-burning campaign against the vital communication and supply link, and the frigid weather added to the misery of the men from Florida. The troops had fought with determined gallantry at the Battle of Perryville, Kentucky. Bragg's timid leadership,

---

[1] Waters and Edmonds, *Small but Spartan Band*, 22–28; Coles, *Florida Civil War Heritage Trail*, 65.

however, snatched defeat from the jaws of victory, and such wasted valor became the trademark for the Confederacy's Western army.[2]

After driving the Federals from Richmond, Gen. Lee decided to take the war into the North. Following an easy victory at Harper's Ferry, site of the capture of abolitionist martyr John Brown, the Rebels and Yankees fought to a bloody stalemate along the banks of Antietam Creek. Following Lee's subsequent withdrawal to Virginia, Pres. Abraham Lincoln used the Federal "victory" as an opportunity to announce the Emancipation Proclamation. He decreed, "that all slaves in regions still in rebellion on 1 January 1863, would be forever free." The president added that on that date, the United States would also begin enlisting Black men into its armed forces. Most Northern military experts predicted that thousands of Negroes would enlist in the Union army to fight for their own freedom—and they were not disappointed.[3]

Several Union officers "jumped the gun," not waiting for the New Year's deadline. Abolitionists Maj. Gen. Rufus Saxton and Gen. David Hunter began recruiting the First South Carolina Infantry, US Colored Troops (or USCT) almost immediately. In late August, Northern newspapers announced "gunboats and steamers have gone to the coast of Florida for [Black] recruits." Florida politicians responded by offering "a large reward for the capture of any…[White] officers [commanding the Black soldiers]." Rather than treating them as prisoners of war, the White leaders of Black troops were to be "hung from the first tree."[4]

As could have been expected, the vast majority of White Southerners viewed Black soldiers with animosity or humorous contempt. Many Federal troops also initially expressed an aversion to Black soldiers fighting for the Union. A Rhode Island journal reported on Maj. Gen. David Hunter's disappointing first attempt to raise a Black regiment:

---

[2] Coles, *Florida Civil War Heritage Trail*, 64. See generally, Sheppard, *Noble Daring*.

[3] Schafer, *Thunder on the River*, 89; For a general account of the Sharpsburg Campaign, see Murfin, *Gleam of Bayonets*.

[4] "The South Carolina Negro Regiment," *Boston Evening Transcript*, 8 November 1862, 4; "Yankees; General Jackson; Florida," *Richmond [VA] Examiner*, 23 August 1862, 2; "From the Florida Coast," *Lowell [MA] Daily Citizen & News*, 30 October 1862, 2; Proctor, *FL100*, 25 August 1862 and 27 August 1862; Joes, *America and Guerrilla Warfare*, 93. See generally, Dobak, *Freedom by the Sword*.

> The Negro Brigade organized by General Hunter has proved an unmitigated failure. Out of eight hundred contrabands on the muster roll, there were some 500 who "skedaddled," and many of the remainder felt so uneasy under military control and discipline they watched for an opportunity to escape. Gen. Hunter...was unable to check opposition to the Negro Brigade which has seized [the White Union soldiers] almost unanimously.[5]

A letter discovered on the person of a dead bluecoat stated, "I went to fight for my flag, and not to fight for the black scoundrels. I am afraid there will be a mutiny in our camps before long." The majority of Union soldiers, however, likely came to agree with an Iowa trooper who wrote, "If any African wants to stand between me and a rebel *bullet* he is *welcome* to the honor and the *bullet too*." The animosity between White and Black bluecoats usually lessened as the war progressed, and a real spirit of comradeship developed as they fought together and endured the dangers and hardships of soldiering. That camaraderie was not the case with White Floridians who enlisted in the Union army. They largely retained their prejudices against the USCT, and that bigotry would seriously hinder Federal operations in the Florida.[6]

The changed status of slaves led many in the North to develop schemes to turn Florida into an abolitionist's utopia. Northern journalists all seemed to believe "the army of the rebellion [in Florida] will be scattered in a few weeks." A Pennsylvania journal, for example, declared, "Florida is virtually in our possession." One abolitionist plan suggested, "President Lincoln contemplates making Florida a cotton plantation: inviting laborers, black and white, to settle there for...[that] purpose, who will have ample protection from both the army and navy...Florida [is] to be reduced to territorial status." Another plan differed in two significant particulars. "[N]orthern and foreign emigration" would furnish the manpower to defeat the Southerners, who selfishly opposed giving up their farms and plantations as part of a Federal social experiment; and the newcomers would arrive "in such formidable bodies [numbers] as to defy the guerrillas." Kansas Sen. Jim Lane—generously described by historians as "mentally unstable" and "eminently unscrupulous"—remained a darling of the abolitionist movement. The Jayhawker proposed

---

[5] Dobak, *Freedom by the Sword*, 9–10.

[6] "From Port Royal," *Providence [RI] Evening Press*, 15 August 1862, 3; "More Northern Sentiment," *Macon [GA] Telegraph*, 20 December 1862, 3.

ending the depredations of the Florida bushwhackers "by setting the slaves…free, and setting them to hunt them [Florida guerrillas] out."[7]

St. Johns River

Gen. Joseph Finegan quickly recognized that the Confederacy could never control the St. Johns River as long as enemy gunboats sailed, with virtual impunity, along the waterway. If the Rebels could halt, or severely restrict, enemy gunboats on the St. Johns, it would permit the Confederacy to "relieve the valley from the marauding incursions of the enemy and…establish a base for operations against St. Augustine." As Daniel Schafer indicated, "He [Finegan] understood that control of the St. Johns River was essential if the interior [of East Florida] was to remain safe." With that in mind, "Old Barney" (as the soldiers called the Irish general) had devised a plan to keep the US fleet bottled up around Mayport Mills. Luckily, by September 1862, he had both the men and materials on hand to put his plan into motion.[8]

Lt. Col. C. F. Hopkins and his 1st Florida Special Battalion had been, until recently, guarding the Apalachicola River to prevent a Yankee incursion into Georgia. When replacement troops arrived to relieve Hopkins's men, Finegan ordered the battalion to proceed to Jacksonville and fortified St. Johns Bluff. The unit's historian described that place as "a good position for placing guns to dispute the passage of enemy gunboats. In what was an otherwise flat country, the bluff rose seventy feet or more above the river with the ship channel passing close to its bank." From the initial occupation of Jacksonville (when Hopkins's troops had burned Union property and received much of the blame for "the night of terror"), the Yankees and Rebels had both recognized the potential of St. Johns Bluff. A Federal sailor wrote, "About four miles above Mayport, on the St. Johns Bluffs (the site of the old Spanish [actually French] Fort Caroline), bold highlands that rise perpendicularly thirty feet from the water, the rebels had cleared away a considerable

---

7 "Items of War News," *Jackson [MI] Citizen*, 1 October 1862, 2; "Colonizing Secessia," *[Worchester] Massachusetts Spy*, 10 September 1862,2; "A Comprehensive View of the Situation," *Philadelphia Inquirer*, 18 November 1862, 4; "Jim Lane's Manifesto," *[Atlanta] Southern Confederacy*, 28 June 1862, 2; Pritchard, *Raiders of the Civil War*, 83; Brownlee, *Gray Ghosts of the Confederacy*, 37; Dobak, *Freedom by the Sword*, 9.

8 Schafer, *Thunder on the River*, 93–94; *ORA*, ser. 1, vol. 14, 122; 494; and 638. See generally, Bearss, "Military Operations on the St. Johns."

space and commenced to erect a battery and barracks for troops. The location is a splendid one, and could easily converted into a miniature Gibraltar."[9]

By July, Finegan had also managed to assemble a sizable number of artillery pieces. Among others, he had gathered two rifled thirty-two–pounders, two eight-inch howitzers, two eight-inch Columbiads, and two twelve-pound rifled cannons borrowed from Capt. R. H. Gamble's Light Artillery. A detachment from the Milton Light Artillery would serve as the Rebels' gunners at St. Johns Bluff. Hopkins's men would provide infantry support for the artillerymen, and during this operation, Capt. Winston Stephens and Capt. W. E. Chambers, both of the 2nd Florida Cavalry, acted as the mobile arm for the men in the fortress.[10]

It did not take the Federals long to discover that the Confederates had occupied the bluff. On 1 September, Capt. Lemuel G. Crane, commanding the gunboat *Uncas*, reported spotting a large number of graycoats at the old Huguenot fortress. To keep the enemy distracted until the Rebels could complete the installation of their artillery, Finegan directed Winston Stephens to create as much havoc as possible near Mayport Mills. The guerrilla captain, initially, succeeded quite well. Acting Master Crane, of the Union gunboat *Uncas*, reported,

> I have been greatly annoyed by the divers bands of guerrillas who are passing through these woods and doing all sorts of mischief. On Saturday night last they attempted to burn Mayport. They succeeded in setting fire to one house, but upon the immediate discovery thereof, I began to shell the woods and dispatched some of my men to the spot where the fire was. We drove the Perpetrators off, having been in due time [able] to put the fire out.[11]

The Union gunboats also took this occasion to shell the home of the Broward Family, described by Crane as "a leader in every unlawful act, and a source of annoyance to every respectable [meaning Unionist] inhabitant [along the northern section of the St. Johns River]."[12]

---

[9] Hillhouse, *Heavy Artillery & Light Infantry*, 36; "Comm. DuPont's Operations," *Philadelphia Press*, 3 April 1862, 1; Schafer, *Thunder on the River*, 92–94.

[10] Hillhouse, *Heavy Artillery & Light Infantry*, 35–36; Schafer, *Thunder on the River*, 105–10.

[11] *ORN*, ser. 1, vol. 13, 301–302.

[12] Ibid.

On the night of 9 September, the USS *Uncas* approached St. Johns Bluff. No shots greeted the warship. A runaway slave named Israel had earlier informed the Union commander that the Rebels had implanted artillery at St. Johns Bluff, but Crane initially found the intelligence suspect. He wrote, "having been deceived by contrabands [before], I had very little confidence in the story." At daybreak, Capt. Joseph Dunham, of the Milton Light Artillery, gave the order to commence shelling the enemy gunboat. The gunners "on the Bluff opened fire, getting off ten rounds before the *Uncas* could reply." One of the Confederate shots hit the ship's magazine but failed to explode. The Federal gunboat *Patroon,* after a long delay caused by the shoals at the mouth of the river, steamed in to join the fray. The fight lasted four hours. The sailors claimed that their accurate fire drove the Rebels from the fort for a few minutes, but they failed to take the stronghold and soon retreated to Mayport. Hopkins's reported his casualties as one killed and eight slightly wounded. (The sole death appears to have been an odd accident. While serving as a spotter for the Rebel artillery, Capt. James H. McRory fell from a tree and impaled himself on his own bayonet.)[13]

Meanwhile, Gen. Finegan pled with Richmond to send reinforcements to aid in the fight at St. Johns Bluff. Finegan correctly guessed the Federals' strategy. They would eventually send in their gunboats to keep the Rebels' attention focused on the river while dispatching infantry overland to take the fortress from the rear. If the Confederate War Department would send him one good infantry regiment, he reported, the Rebels might block the Yankees' use of the St. Johns and retake—and hold—St. Augustine. However, the Irish General warned that if help did not arrive quickly, they could not hope to hold the Florida Gibraltar. As the threat of a major Federal expedition against St. Johns Bluff increased, so did "Old Barney's" pleas for assistance.[14]

On 16 September, the Yankees made another attempt to take the Bluff. Five Union gunboats pounded the fortress for several hours but could not permanently dislodge Hopkins's stubborn Confederates. Despite the barrage

---

[13] Hillhouse, *Heavy Artillery & Light Infantry*, 36–37; *ORN*, ser. 1, vol. 13, 324–25; Schafer, *Thunder on the River*, 106–108. The Broward family were dedicated Confederates with several sons in the Confederate Army. Their youngest son, Washington, led a guerrilla band on the northern side of the St. Johns. In the early 1900s, one of the Broward's sons, Napoleon B. Broward, would be elected governor of Florida.

[14] *ORA*, ser. 1, vol. 1, 14, 121; Schafer, *Thunder on the River*, 107–109.

of shells, neither side inflicted much damage on their opponents. Com. Charles Steedman, commanding the US fleet, reported "the rebels suffered considerably from the incessant shower of shells and projectiles." Despite that optimistic estimate, Hopkins's losses totaled only two killed and four wounded.[15]

The failed naval assault convinced Com. Steedman that only a combined operation—using both infantry and naval forces—could dislodge the Rebels from their fortress. By 1 October, Steedman had assembled an infantry force, estimated from sixteen hundred to three thousand men, and seven gunboats. Finegan, whose pleas for reinforcements were ignored, left Tallahassee for Jacksonville. Lt. Col. Hopkins would have a combined artillery, infantry, and cavalry force of about seven hundred men to oppose the coming Union offensive.[16]

Early on the morning of 2 October, the Union ships began unloading infantry troops at Pablo Creek, east of St. Johns Bluff. Because of the marshy terrain, the bluecoats would have to march south to the stream's headwaters before turning west to attack the Rebels. Chambers promptly informed Hopkins that a second enemy contingent had also begun debarking at Greenfield's Plantation. Chambers did what he could to delay their advance. Spreading his forces through the heavy undergrowth, the guerrilla chieftain and his men began sniping at the enemy. Like an overmatched boxer, Chambers tried to delay his opponents with feints and quick jabs. When pushed, Chambers's irregulars would retreat, reform, and start the whole process again. His hit-and-run tactics delayed the Federals briefly but could not halt the inevitable.[17]

Maj. T. W. Brevard, commanding a sizeable Confederate infantry unit in the area, requested permission from Hopkins to attack the Yankees, but he leader refused. Hopkins reasoned that with two enemy columns in the area, if Brevard attacked one, the second contingent would almost certainly strike the Rebel infantry in the flank or rear. Between the difficulties of unloading

[15] *ORN*, ser. 1, vol. 13, 329–30, and 362–65; Schafer, *Thunder on the River*, 109.

[16] Hillhouse, *Heavy Artillery & Light Infantry*, 38–43; Schafer, *Thunder on the River*, 109–13.

[17] *ORN*, ser. 1, vol. 14, 138–41; Hillhouse, *Heavy Artillery & Light Infantry*, 40–44; Schafer, *Thunder on the River*, 110–18.

the troops, and the annoying guerrillas, the Union foot soldiers did not begin the march toward the Bluff until 1 p.m.[18]

Once the bluecoats got started, they advanced so quickly that Chambers felt compelled to burn his own camp rather than letting it fall into enemy hands. With all chance of seriously delaying the Federal advance gone, Hopkins directed Chambers to fall back to Maj. T. W. Brevard's position. If the situation were not bad enough for the graycoats, a hard rain began falling, making much of the Confederate ammunition useless. Hopkins held a counsel of war to discuss the situation and quickly decided their "forces there [at St. Johns Bluff] were insufficient to repel the enemy in front." By 9 p.m., the Rebels had evacuated their stronghold on the St. Johns River. In his haste to get his men out of harm's way, Hopkins left without spiking his artillery or removing the powder, shot, or shells. The commander tried to minimize his culpability, noting that he had nothing with which to spike the guns, but failure to remove the valuable munitions seems to indicate a measure of panic in the Southern ranks.[19]

In the aftermath of the engagement at St. Johns Bluff, the Federals again briefly occupied Jacksonville. Their gunboats sailed up and down the St. Johns, destroying all the small vessels they could find, razing Rebel property, and evacuating the Black men (and their families) who had served the Union as pilots or had provided information. The bluecoats particularly wanted to locate the *Gov. Milton*, the gunboat that had carried the military supplies to the Confederates at St. Johns Bluff. The Unionists eventually found the steamer near the village of Hawkinsville and soon put it to work in the Federal Navy. Com. Maxwell Woodhull, on observing the vast potential he saw in the fertile region along the St. Johns, attempted to convince his superiors of the wisdom of permanently occupying Jacksonville. He wrote, "The cattle of Georgia, Alabama, [and] North and South Carolina have all been consumed. Texas and the rich grazing country to the westward of the Mississippi [are] being cut off, the whole dependence of the Confederate Government to feed their Army now rests on this State." His prescient observation was ignored in Washington, and the failure to act upon his suggestion would cost the Yankees dearly for much of the remainder of the war.[20]

---

[18] Schafer, *Thunder on the River*, 107–109.

[19] Ibid.

[20] Ibid.

The evacuation of St. Johns Bluff infuriated Finegan. The retreat left Jacksonville and all of Northeast Florida wide open to the bluecoats. The Irish general brought Hopkins up on charges, but the commander of the 1st Florida Special Battalion had the perfect defense. He quoted the pleas "Old Barney" sent to the War Department in Richmond. The Irish general's requests for reinforcements had stated that if the Rebels at St. Johns Bluff did not receive reinforcements, the position could not be held. Therefore, Hopkins reasoned, since he had received no additional troops or weapons, he could not be blamed for abandoning the fort. The court agreed with Hopkins's argument. Not one to be stymied by a setback, Finegan decided to reorganize the battalion into a regiment. Then, he could name his adjutant, Robert B. Thomas, to lead the unit. "I am anxious," Finegan explained, "that this regiment should be organized as proposed, being satisfied that its efficiency will be much increased by having at its head an officer of some experience in the army, and particularly with foot soldiers." By comparing Thomas, the West Point graduate, and Hopkins, who had attended the Naval Academy, he apparently thought that he could sway the officers and politicians in Tallahassee and Richmond to support his candidate. Hopkins, however, again prevailed and retained command of the unit. With that failed scheme, Finegan found another way to exact his revenge and banished Hopkins and the 1st Florida Special Battalion to Thunderbolt, Georgia. There they would languish in that marshy backwater, near Savannah, for more than a year.[21]

TAMPA AND SOUTH FLORIDA

The situation at Tampa seemed, on the surface, to be progressing well for the Confederates. Capt. J. W. Pearson remained at Fort Brooke, and the area's Rebels managed to foil most Federal operations in the Bay area. In late August, three Unionist refugees, who had been living on Egmont Key for several months, decided to slip back to the mainland "to procure potatoes, beef, etc. from their farms near Old Tampa, for the support of themselves and their families." A band, described only as "guerrillas," ambushed the group, killing one man, capturing another, and mortally wounding John Whitehurst, the leader of the Unionist raiders. The partisan commander, who directed the ambush, appears to have been Abel Miranda. (Rumors had circulated through the region for months that the Whitehurst brothers had instigated

[21] Proctor, *FL100*, 29 August 1862; *ORN*, ser. 1, vol. 17, 309.

the shelling of the Miranda homestead, and the bushwhacker wanted nothing more than to exact his revenge on Whitehurst.) Badly wounded, the Unionist managed to escape the partisans' trap but subsequently suffered the agonies of the damned as he floated in the Gulf in an open boat under a blazing sun. Federal sailors found Whitehurst just before he died, but Miranda's brutal actions served as a warning to the Unionists in South Florida—enter Rebel territory at your own risk.[22]

In late October, frustrated US Naval authorities decided to take a page from the guerrillas' playbook. Pearson described the incident:

> We had one of the most rascally attacks from the Yankees, upon this place [Tampa], you ever heard of being made by a civilized people. A vessel came up yesterday in disguise—showed no colors even after I raised mine—tried to make [us] believe it was a vessel running the blockade...but as soon as she got in [position to attack] about seven o'clock, A. M., [it] opened fire upon the town with shot and shell among women and children. Nobody killed, several hairbreadth escapes, doing damage to the town, etc.... [They]...kept up their fire until about ten o'clock, [but] they could affect nothing at this. We replied to them: The game seems to be now—If you don't *run, we will.* So we stood our own ground and they *run.*[23]

Despite his military successes, many citizens of Southwestern Florida quickly tired of Pearson and his overly zealous actions. A substantial number of people in that region had little love for the Confederacy, and the Conscription Act found even fewer supporters. These included people born in the North, who had moved to Florida before the conflict; the foreign-born, who almost without exception loathed slavery; and poor Whites, to whom "the loss of a few head of stock [cattle], or the failure of a five acre crop meant temporary ruin." Pearson, of course, fervently pursued his duty and "remorselessly pressed [reluctant men] into the Rebel army." T. J. C. Howell, of the gunboat *Tahoma*, stated, "These guerrillas are scouring the woods, looking after deserters and conscripts; they rob, murder, and steal indiscriminately, if reports of refugees are to be credited; Union men they threaten to hang and

---

[22] Hillhouse, *Heavy Artillery & Light Infantry*, 47–48; *ORA*, ser. 1, vol. 14, 665; *ORA*, ser. 1, vol. 52, pt. 2, 350.

[23] Proctor, *FL100*, 29 August 1862; *ORN*, ser. 1, vol. 17, 309.

[they] do shoot, as we have lamentable proof. It has been said that every man capable bearing arms has been forced to join the rebels in this part of Florida."[24]

Even Southern sympathizers sent complaints to Tallahassee regarding Pearson's troops. The exact nature of the charges leveled against his "Oklawaha Rangers" is not revealed in extant records, but they probably involved the taking of food to supplement the soldiers' meager rations. Gen. Finegan castigated Pearson, but there is no record of official action being taken against the guerrilla or his men.[25]

As if the region did not have problems enough, "sometime before the end of the year [1862] came an outbreak of smallpox so violent [virulent], at its height, that the congregation of the Peas Creek Baptist Church was afraid to even meet for prayer." Historian Canter Brown Jr. lists at least ten elderly residents who died during this epidemic. Adding to the region's woes, the Seminoles had become fearful when traders and merchants refused to sell gunpowder to tribe members. Thinking it might represent a renewal of hostilities, a delegation of Native People went to Tampa to discuss the situation. The Whites explained the war raging throughout the South had made obtaining ammunition for the Rebel army difficult. With their anxiety put to rest, the Seminole chief dictated a letter to Gen. Beauregard "offering to become allies of the Confederacy." During this time, the Confederate Congress "came to the rescue. Reflecting the increased dependence of the Confederate armies on [Florida] beef, the Congress on 11 October 1862, granted an exemption from the draft for [the owners of] each 'five hundred head of cattle.'" With over sixty thousand head of cattle in Polk and Manatee Counties, this proved extremely beneficial for a number of cattlemen and their drovers.[26]

CEDAR KEYS

Modern readers, raised in an age of refrigeration, often find it difficult to appreciate the importance of salt to the Civil War generation. Roman soldiers had received their wages in salt (or *salarium*, from which we get our word

---

[24] "Particulars from Tampa," *Charleston [SC] Courier*, 13 November 1862, 2. The date of Pearson's letter was 26 October 1862.

[25] *ORN*, ser. 1, vol. 17, 309; Waters, "Tampa's Forgotten Defenders," 13–14.

[26] Waters, "Tampa's Forgotten Defenders," 14; "An Act to Exempt certain persons from military service," 77–78; C. Brown Jr., *Florida's Peace River Frontier*, 152–53; "The Seminoles," *Macon [GA] Telegraph*, 18 November 1862.

"salary"), and for centuries people used salt to preserve meat and fish and for tanning leather. By mid-1862, Floridians by the hundreds had moved into the lucrative business of processing seawater into salt. "The Confederacy needed six million bushels of salt annually," and the Gulf Coast ecology provided virtually everything the salt-makers needed. Nature provided saline water; pinewood for the fire used to boil the liquid; and miles and miles of scrub to conceal their activities. Additionally, salt workers received an exemption from Confederate military service. More importantly, a salt maker could earn a small fortune for a few months of work. An estimated twenty-five hundred men worked at St. Andrews Bay, the site of the southernmost state's largest salt-making facility. The owner of several kettles there, observed, "The destruction of the salt industry in [at] St. Andrews Bay would be a greater blow [to the Confederacy] and more severely felt [by its citizenry] than the falling [capture] of Charleston."[27]

By the end of 1862, various ships of the East Gulf Blockading Squadron began conducting "harassing raids" on the salt makers and their works. As naval historian George E. Buker explained, "These expeditions were not a concerted effort to wipe out Florida's saltmaking capabilities.... They were made by individual commanding officers to relieve the monotony of blockading duty [rather] than to strike a telling blow to the Confederacy." It became almost a game. The Union blockader hove into view, and the salt-makers, both White and Black, scurried into the scrub, and the sailors and marines came ashore and destroyed the pots and any bags of salt foolishly left behind. There may have been a few random shots, but the workers generally just let the sailors have their fun. After destroying the facilities, the Unionists would return to their ships and as soon as they vanished over the horizon, the salt-makers returned, replaced their kettles, and resumed their lucrative undertaking.[28]

On 4 October 1862, the "game" took a more serious turn at Cedar Keys. The USS *Sagamore* threw a few shells at a distillery on Depot Key (one of the Cedar Keys). The shelling continued until three white flags appeared in the windows of a nearby dwelling. A contingent of sailors and marines rowed to land and moved toward the house flying the white flags. A newspaper

---

[27] Ekardt, "The Navy's Great Salt Raids"; Coles, *Florida Civil War Heritage Trail*, 9–10; Buker, *Blockaders, Refugees, & Contrabands*, 46.

[28] Ekardt, "The Navy's Great Salt Raids"; Buker, *Blockaders, Refugees, & Contrabands*, 51.

reported, "It seems that a guerrilla band enticed a boat ashore by displaying three white flags from the houses on shore. As soon as the men had landed they were fired upon from the windows of the houses: the flags of truce were flying the whole time. The sailors got off [escaped] with their wounded...."[29]

The *Somerset* signaled for reinforcements, and the *Tahoma* soon arrived. According to a modern source, "A couple of days later they [the Federal blockaders] landed a larger force and destroyed some larger [salt] works in the vicinity. There were two extra heavy cast iron kettles and two wrought of [from an] iron boiler that were too thick to break up with sledgehammers, so the sailors put a couple of howitzer shells through them." The captain of the *Somerset* declared, "the Rebels needed a lesson and they have had it." Several newspapers reported the Federals "burned the town," but it appears they only burned the houses used by the bushwhackers and a couple of nearby buildings.[30]

The identity of the "guerrillas," who decoyed the sailors into the trap remains unknown. Capt. Edward J. Lutterloh commanded the only known organized Rebel squad operating in the area. Lutterloh, who later directed a "Cow Cavalry" unit in Levy County, had been a friend and employee of former Sen. David Yulee and often aided salt makers near Cedar Keys. The island's residents had even "immortalized" the captain by naming "a body of water used, under cover of darkness to deliver supplies to men working [at] the salt-making facility" Lutterloh's Ditch. Whoever shot up the Federal sailors, Cedar Keys had not seen the last of the Federals.[31]

NASSAU COUNTY AND SOUTH GEORGIA

Nowhere did the Emancipation Proclamation have a more immediate and profound effect than in the Department of the South. There had been a few missteps in enlisting Black troops, as when Maj. Gen. David Hunter's attempted to form his First SC Regiment (USCT) at the point of a bayonet, but there were plenty of volunteers among the free Blacks and former slaves who had fled the plantation for freedom. By September 1862, Black units began conducting raids in the Nassau County and South Georgia

---

[29] "An Expedition to Apalachicola," *Philadelphia Public Ledger*, 1 November 1862, 1; *ORN*, ser. 1, vol. 17, 316–19.

[30] Ekardt, "The Navy's Great Salt Raids"; "An Expedition to Apalachicola."

[31] Fishburne Jr., *The Cedar Keys in the 19th Century*, 75–76; McCarthy, *Cedar Key, Florida*, 29.

hinterlands. Each expedition increased the number of African Americans who were willing to fight for freedom. Lieutenant Colonel Oliver Beard reported, "I started from St. Simon's with 62 fighting men (all colored). As soon as we took a slave from his claimant [master] we placed a musket in his hand and he began to fight for the freedom of others."[32]

The military situation in Nassau County during the final months of 1862 could not have seemed more lopsided. The Federals had a firm foothold on Amelia Island. They maintained one full White regiment at Fernandina, as well as a sizeable number of Black troops—loosely attached to the First South Carolina Colored Troops (USCT)—who had recently been posted at Fort Clinch. The Federals had also established a Unionist haven on Batton Island, near the mouth of the St. Johns River, to house "a colony of white refugees...[who were] all Union men of Southern birth who fled from their homes to escape Southern conscription." The loyalist community there beseeched the Federal officers at Fernandina to burn the bridge at Batton Island "to keep the guerrillas away, whom those people so much dread." Capt. Frank Clarke's Rebel cavalry, numbering about eighty men, remained the lone Confederate unit in the area to oppose the large number of occupation troops.[33]

An incident occurred on 26 October that caused outrage throughout the Confederacy. On that day, the Union Navy destroyed St. Marys, Georgia, a beautiful hamlet located just across the Florida state line. Reporting the destruction of the picturesque village, the *Southern Confederacy* noted, "Two Yankee gunboats destroyed the town of St. Mary's on Sunday last, by throwing shell and hot shot [from their vessels]. The town is entirely destroyed. The Yankees attempted to land at St. Mary's, but were repulsed by our pickets. They then destroyed the town." Ignoring the presence of skirmishers near the village, the Rebel press deplored the "overkill" used on the beautiful little community. The hamlet had little strategic value and posed no real military threat to the Federals. The Southern press pointed out that the sole occupants were three or four elderly women. Confederate citizens and the press pointed to this incident, whether justified or not, as an example of "Yankee villainy."[34]

---

[32] Schafer, "Freedom Was as Close as the River," 169–70; Glatthaar, *Forged in Battle*, 6; Dobak, *Freedom by the Sword*, 35–37.

[33] "A Visit to Jacksonville," *Augusta [GA] Daily Constitutionalist*, 13 December 1862, 1; *ORN*, ser. 1, vol. 13, 245.

[34] *Atlanta Southern Confederacy*, 15 November 1862, 3; "Vandalism in Georgia and Florida," *Fayetteville Carolina Observer*, 17 November 1862, 4.

With the Confederates occupied by threats to the St. Johns River Valley, the Unionists at Fernandina used this as an opportunity to conduct a series of raids throughout the Nassau County and South Georgia. On 3 November, Capt. Charles Trowbridge and "a company of colored troops" began by destroying salt works along the Bell and Jolly Rivers and burned the Confederate encampment at Camp Cooper. The next day they laid waste to the salt-making operations at King's Bay, Georgia. As Trowbridge and his men headed back to the *Darlington*, Clarke's Rebel horsemen ambushed the bluecoats. A Union account stated: "They [the Confederates] had in all about 50 [men] in number, [and the Black troops] had just got[ten] into small boats, when the enemy rushed out of the thick woods, and fired upon them. Their [the Black soldiers'] condition was a perilous one, the enemy not being ten rods distant and the steamer still further off. Nothing daunted, the men [Black troops] loaded and fired coolly and incessantly, till safe aboard."[35]

Union officers often felt compelled to portray African American soldiers as stalwart warriors to the people of the North, and that sometimes led to some exaggeration in their reporting. Raw troops caught by surprise, in open boats and receiving heavy fire, must have nerves of steel to exhibit no panic. Southern journals, with their own set of prejudices, asserted, "The black troops are exceedingly awkward [in combat], and are handled roughly by their white officers."[36]

ST. AUGUSTINE

The Ancient City remained a city under siege throughout the remainder of 1862, but political maneuvering made the headlines during this period. Lt. Col. Louis Bell, of the Fourth New Hampshire, became an early victim of the abolitionist agenda. Sec. of War Edwin M. Stanton had appointed Greenfield, Massachusetts, native Gen. Rufus Saxton, a devout believer in the wickedness of slavery and 1849 graduate of West Point, as commander of the Department of the South. Stanton had specifically directed Saxton to do everything in his power to end servitude in his district. The Union troops at St. Augustine quickly formed an intense hatred for Saxton and the coterie of

---

[35] "The War in Florida," *Pomeroy [OH] Weekly Telegraph*, 28 November 1862, 2; "A Visit to Jacksonville"; "A Brief History of Clinch's Regiment, 4th Georgia Volunteer Cavalry." Camp Cooper appears to have been located near the present day village of Yulee.

[36] "The War in Florida"; "A Visit to Jacksonville."

toadies who followed in his wake. A Federal officer, commenting on Saxton's initial visit to the Ancient City, stated,

> Gen. Saxton...came down with a party of his civilian assistants who were called "Gideonites" by the soldiers. Those people were almost generally shirks and sneaks destitute of [the] courage to fight, but eagerly follow in the wake of invading armies to collect up the abandoned property.... It was only by the utmost vigilance on my part that both [Mansfield] French [a Gideonite and Methodist minister] and Saxton were saved from being assaulted and roughly handled by the incensed soldiers.

William A. Dobak, in his excellent history of the USCT, suggests that the true reason for the animosity ran deeper than cowardice and profiteering. "Unlike the Gideonites," he wrote, "[the] Northern soldiers had not come South to free slaves, but to crush the rebellion."[37]

Saxton's first priority was the removal of Col. Bell as commander of the Fourth New Hampshire at St. Augustine. "It is with deep regret that the General Commanding the Department has received several reports against officers for returning fugitive slaves," he wrote, "in direct violation of a law of Congress...[Bell has] engaged in the manly task of turning over a young woman, whose skin was almost as white as his own, to the cruel lash of her rebel master." The men of the Fourth New Hampshire felt particularly incensed, feeling Bell had been "publically declared guilty in [a] general order, before...[convening a] trial." In Bell's absence, Lt. Col. Gilman E. Sleeper, a prewar cobbler, directed the Granite State troops.[38]

As it turned out, Col. Bell had a legitimate reason for his supposedly heartless action. The US surgeon at St. Augustine reported that the town was "infested" with prostitutes. These "camp followers" invariably showed up in the wake of an occupying army, and incidences of venereal diseases increased dramatically. Bell admitted that he "sat them [including the light-skinned mulatto woman] outside the city gate, whether they were black or white."

---

[37] Schmidt, *Florida's East Coast*, 293; Dobak, *Freedom by the Sword*, 28.

[38] Schmidt, *Florida's East Coast*, 294; "Our Army Correspondence," *Amherst [NH] Farmer's Cabinet*, 2 October 1862, 1.

(Brig. Gen. John M. Brannan would restore Col. Bell to command in September 1862.)[39]

Gen. Saxton's next edict elicited screams of outrage throughout the Confederacy. Those die-hard Rebel women gloried in their obdurate attitudes, and they went out of their way to show their disdain for the occupying force. Gen. Saxton decided to get rid of the problem once and for all. He ordered all White citizens in St. Augustine to assemble at the Presbyterian Church. One of the "Gideonite" spokesmen began the meeting by comparing the town's Southern women to streetwalkers on New York City's Broadway. He then demanded all of them to sign a loyalty oath or suffer banishment. Two days later, forty-eight women and seventy-six children were herded aboard the steamer *Burnside* and began the journey up the St. Johns River. The women and children were given no food or water, and the Federals even refused to provide liquids when some of the refugees became seasick. Kept below deck, the Union sailors delighted in informing the frightened women that "the negroes had attempted to scuttle the vessel in order to drown the Secesh [women and children]." When the steamer reached Jacksonville, Gen. A. H. Terry came aboard the vessel and revoked the order of exile. He directed the *Burnside* to return to St. Augustine, "where the women and children were set at liberty." Since they had been forced to sell all their property before exile, most "were quite destitute." The damage had already been done. The idea of making war on women and children incensed most Southerners and strengthened their resolve to continue the fight.[40]

During the final months of 1862, the situation changed dramatically at St. Augustine. The Seventh New Hampshire replaced Col. Bell and the Fourth New Hampshire, and the new unit brought with them an entirely different attitude. A member of the newly arrived unit, Calvin Shedd, reported, "The Citizens [of St. Augustine] liked the 4th [New Hampshire] first rate for they used to abuse the [Black people] & let the whites do as they pleased, but when the 7th [Regiment] came here everything changed." It did not take Shedd long to comprehend the real situation at St. Augustine. "I don't think there is a real hearty Union man in Town, [but] they take the

---

[39] "St. Augustine," *Manchester [NH] Daily Mirror*, 19 September 1862, 2; "Col. Louis Bell."

[40] "St. Augustine Affairs," *Philadelphia Inquirer*, 8 August 1862, 1; "Yankee Atrocities in Florida," *Charleston [SC] Mercury*, 15 October 1862, 1; Graham, "Home Front," 31–33.

oath for convenience sake, so they can stay at home." The new Granite State unit "abandoned conciliation for 'hard war.' Hard war…is a specific form of warfare that strategically targets not just military personnel and material, but also civilian property thereby denying [those] materials for the enemy army."[41]

Despite the various changes, the Rebels' guerrilla campaign continued against the Ancient City occupiers. Complaints of lack of food and fuel, caused by the presence of Confederate irregulars, are found repeatedly throughout the Union correspondence during this time. On 16 August, Capt. John Westcott, commanding a company in Maj. T. W. Brevard's Partisan Ranger Battalion, captured two men who had been previously warned to stop selling beef to the Yankees. "The two turncoats, and a man named Shoemaker, who had previously deserted the Confederate army, were killed within a half-mile of Fort Marion. Skirmishes continued to be an almost weekly occurrence, and three companies of Union infantry had to accompany woodcutters whenever they went out for fuel." As Omega G. East noted, "In truth, action in St. Johns County involved merely minor skirmishes, which served the useful purpose of keeping the Federal garrison pinned down, and discouraged looting expeditions into the interior of the state."[42]

St. Johns River

News of the Emancipation Proclamation spread among the slaves of North Florida. Several Negroes worked for the US troops at St. Augustine, with wages paid to their masters, but Gen. Saxton chose to ignore Pres. Lincoln's timeline for Emancipation. In October 1862, Saxton assembled all the town's Black people and explained to them that they were now free. A New Hampshire soldiers reported, "This made them wild with joy…. It seemed the [biblical] Day of Jubilee to them." Those who returned to the plantations quickly spread the wondrous news. The results were predictable. All "along the South Atlantic Coast that could be reached by boats from Union warships were abandoned plantations, deserted towns, wasted crops, stray cattle, and wandering slaves." Dr. Schafer further noted, "Waving white flags from shore, paddling canoes and poling rafts, slaves and free blacks hailed Union vessels.

---

[41] Calvin Shedd to S. Augusta Shedd, 12 October 1862, quoted in Totten, "Ancient City Occupied," 57–60.

[42] Schmidt, *Florida's East Coast*, 297, 298, 305, and 310; East, "St. Augustine during the Civil War," 89.

For the enslaved men and women of northeast Florida who could escape to the Union gunboats, the possibility of freedom was suddenly as close as the St. Johns River."[43]

The Confederate authorities realized that they must do something to stop the flight of their Negroes. Gen. Finegan wired Capt. J. J. Dickison with directions to "remove all negroes having no owners with them and free negroes from the Saint Johns River into the interior a safe distance from the enemy, and place them in charge of some white person, to be held subject to their owners' orders, and in the case of free negroes to be left in their own charge, subject to the laws of the state."[44]

Dickison headed east, planning to operate in the area around Palatka. In his postwar account of the conflict in Florida, the guerrilla leader (who seemingly viewed this operation with distaste) glossed over the events during this campaign. "After the enemy began demonstrations on the St. Johns," he wrote, "the command was ordered to Palatka, 75 miles from Jacksonville. While on the march they captured a large number of negroes who were endeavoring to escape to the enemy, and by this timely capture discovered a plot which had been set on foot to drain that part of the country of slaves. They also captured a number of deserters." As the depth of the conspiracy became apparent to the politicians and military leaders, Gen. Finegan ordered Capt. W. E. Chambers and his unit to reinforce Dickison's company.[45]

MAGNOLIA SPRINGS (CLAY COUNTY)

One final—rather odd—episode played out along the St. Johns River before 1862 ended. It began with a letter from Com. Maxwell Woodhull to Acting Master William Watson, commanding the USS *Uncas*. Woodhull requested Watson to check on Rebel activities around Black Creek, in Clay County. Woodhull wrote, "I understand that the guerrilla company of Capt. John Westcott is located near that point and is now in a state of disorganization. If this is true, you may be able to pick up some fugitives from [this] unwilling service, [and] also some of their unfortunate black brethren." He also

---

[43] "Emancipation," *Milwaukee Sentinel*, 28 October 1861, 1; "The Civil War Fifty Years Ago Today," *Lexington Leader*, 9 March 1912, 9; Schafer, *Thunder on the River*, 84.

[44] *ORA*, ser. 1, vol. 14, 661.

[45] Ibid.; Evans, ed. (and J. J. Dickison), *Confederate Military History* (Florida), 11:53.

specifically directed the *Uncas* to stop to check on Dr. Nathan Benedict, a known Union man and owner of a large hotel at Magnolia Springs. Earlier Benedict had advised Woodhull that he wanted to send his son to the North to avoid his conscription into the Rebel army.[46]

On either 5 or 6 October, the *Uncas* anchored opposite Dr. Benedict's hotel and sent a launch ashore to check on the "Union" man. Benedict met the Federal sailors at the end of his long wharf, chatting with the Yankee tars as they headed back to the shoreline. Upon reaching dry ground, the innkeeper suddenly jumped off the dock and concealed himself under the plank walkway. Almost simultaneously, a volley of rifle fire came from the nearby woods, which badly frightened the bluecoats but inflicted no casualties. The Federal sailors made a mad dash toward their vessel—and all managed to reach the gunboat safely.[47]

Watson wrote, "The men at once retreated back in good order to the boat, exposed to a continuous fire from the…enemy." After that, "the *Uncas* promptly and effectively shelled the locality." Apparently, neither the Confederates nor Federals suffered any casualties from the ambush. Watson blamed the attack on Westcott and his company. The acting master reported his vessel became stuck on a sandbar that prevented him from getting into a position to destroy Benedict's hotel.[48]

It appears obvious from the tone of Watson's official report that he believed Benedict facilitated, if not planned, the ambush. The hotelier, however, had unblemished credentials as a "tried and true" Union man, so it leaves open the question of why he would cooperate with Westcott's men. Perhaps the Rebels threatened to burn his property, kill him, or hold his son hostage to insure his collaboration. If, however, Benedict actually aided the Confederate of his own free will, no reason for his duplicity has come to light.[49]

---

[46] *ORN*, ser. 1, vol. 13, 464–65; Schafer, *Thunder on the River*, 35.

[47] *ORN*, ser. 1, vol. 13, 469–70.

[48] Ibid.

[49] Ibid.

Chapter 4

"Charge in among Them!"

(January–July 1863)

Fernandina and Nassau County

The initial Union operation of 1863 in Florida began with an attempt to seize a large amount of cut and seasoned lumber in Nassau County. The US Navy needed vast quantities of wood to repair and construct the vessels used in blockading duty and for the riverine warfare in the Deep South. The ships *Cosmopolitan*, *Delaware*, and *Uncas* left South Carolina, on 3 January, bound for Amelia Island. The infantry troops for this expedition consisted of about two hundred members of the Third New Hampshire and three hundred men from the Third Rhode Island regiments. The men and vessels reached Fernandina on 5 January.[1]

The Federals wasted no time in getting the mission under way—and immediately ran into trouble. On 6 January, the three vessels reached the mouth of the St. Marys River, crossed the bar, and proceeded up the waterway toward Nassau Mills. Acting master of the *Uncas*, William Watson, wrote,

> While proceeding up the river, and about 6 miles inside the bar, we grounded on a sand bar, where the *Uncas* lay, straining the vessel badly. [We] discovered the same night (while aground) a large fire in the direction of the Nassau Mills, which afterwards proved to be the fire that destroyed the lumber we were after. On arriving at the mills [we] found the two mills standing safely, also a huge pile of cinders still throwing up large volumes of smoke and flames.[2]

---

[1] Schmidt, *Florida's East Coast*, 464; "Our South Carolina Correspondence," *Philadelphia Inquirer*, 14 January 1863, 2; *ORN*, ser. 1, vol. 13, 502.

[2] Schmidt, *Florida's East Coast*, 467; *ORN*, ser. 1, vol. 13, 502; "Highly Interesting from South Carolina," *Philadelphia Inquirer*, 29 January 1863, 2; "Nassau River," *Charleston [SC] Mercury*, 27 January 1863, 1; "Florida 2nd Cavalry, Co. K."

On the way back from their unsuccessful mission, the Federals again ran aground on the same sandbar. While they attempted to free the ship, Watson reported, "The transports were fired upon by a guerrilla band on shore, and three men of the Third New Hampshire Volunteers were seriously wounded. The *Uncas*, however, by a few well-directed shells, and by musketry, quickly silenced the Rebels, inflicting mortal injuries upon such as chose to reveal themselves upon the shore." The partisans likely belonged to Capt. Clarke's company of the 2nd Florida Cavalry, but extant records do not reveal the unit suffering any casualties on 8 January. Another account simply stated, "[T]wo rebels were seen to fall." The Northern press certainly had no copyright on outlandish claims, and a Southern newspaper breathlessly reported, "[T]hirty of the Yankees [were] killed. No loss on our side."[3]

The original aim of the expedition had been the confiscation of hardwood timber for use by the US Navy, but the destruction of the Nassau County stacks had frustrated their plan. Not to be denied, however, "the *Cosmopolitan* subsequently proceeded up the St. Johns River, [and] secured a cargo of lumber." With its mission now accomplished, the flotilla headed north, arriving at Hilton Head on 12 January.[4]

Two weeks later, the war in Florida changed dramatically. In mid-January, approximately five hundred Black soldiers of the First South Carolina Volunteer Infantry (USCT) sailed from Beaufort to Fernandina aboard the USS *John Adams*, *Planter*, and *Ben De Ford*. Col. Thomas W. Higginson commanded the troops. Higginson, a native of Cambridge, Massachusetts, graduate of Harvard, and Unitarian minister had long been an advocate of freeing all slaves. (He had reportedly been one of the "Secret Six" who furnished funds to abolitionist martyr John Brown.) Gen. Rufus Saxton, another devout opponent of slavery, had enlarged the First South Carolina from a single company to battalion strength. Armed and drilled, Saxton and Higginson decided to test them in battle in Nassau County. Dr. Dobak suggested the two officers choose remote North Florida for their test run because "as

---

[3] "A Contraband Expedition Whips the Rebel Chivalry," *Nashville Daily Union*, 17 February 1863, 2 (contains Higginson's official report, with comments); "Thomas Wentworth Higginson"; Dobak, *Freedom by the Sword*, 31–36; See generally, Wilson, "In the Shadow of John Brown," 306–35.

[4] Schmidt, *Florida's East Coast*, 477, quoting Higginson, *Army Life in a Black Regiment*.

the least populous state in the Confederacy, Florida remained an afterthought of [the] Federal military throughout the war."[5]

To his credit, Higginson tried to acquaint himself with the area before taking his men into the Nassau County hinterland. Unfortunately, after the disastrous attempt to capture the *Hard Times*, the US Navy normally gave the St. Marys River a wide berth. As the abolitionist leader wrote, "The last trip up the St. Mary's had been undertaken by the gunboat *Ottawa*, which had to fight its 'way past batteries at every bluff.' I was warned that no resistance would be offered to the ascent [of the St. Marys], but only on the return…. [The Confederates] had dug rifle pits [along the riverbank, to fire at Union vessels.]" Despite the danger, Higginson had total confidence in his Black soldiers.[6]

Almost immediately after arriving at Amelia Island, Higginson loaded six companies of his men aboard three steamers and began the trip up the St. Marys. Higginson wrote, "[W]e went up the St. Mary's river, in the evening and landed at Township, on the Florida side, about twenty miles up [from Fernandina], with the purpose of capturing Captain Clarks' [Clarke's] company of guerrillas, known to be encamped about four miles from the river." The abolitionist pinned his hopes for success on a Sgt. Robert Sutton, "who was thoroughly acquainted with the country." A surprise attack by untried troops, deep in enemy territory, on a moonlit night, might seem the action of a rash, reckless officer, but Higginson apparently had complete confidence in his men.[7]

Before coming to grips with the Rebel guerrillas, Col. Higginson further complicated the battle plan by requiring his untested soldiers to perform a double envelopment—in the dark. (A double envelopment called for one third of the force to advance to the left; one third would go to the right, trapping the Rebels between them. They would then drive the enemy force forward, where the remainder of the battalion would block the enemy's panicky retreat and destroy them.) Marching through a pine forest, even on a moonlit night, required frequent stops to maintain troop alignment. The operation, however, did not go as planned. "There was a trampling of feet among the advanced guard as they came confusedly to a halt," Higginson

[5] "A Contraband Expedition Whips the Rebel Chivalry"; *ORA*, ser. 1, vol. 14, 195–98.

[6] *ORA*, ser. 1, vol. 14, 195–98; Schmidt, *Florida's East Coast*, 480–81.

[7] Schmidt, *Florida's East Coast*, 480–81.

wrote, "and almost at the same instant a more ominous sound, as of galloping horses in the path before us.... The [Confederate] leader of an approaching party mounted on a white horse and reigning [reining] up in the pathway...drew a pistol from the holster and took aim; others [USCT] soldiers heard the words 'Charge in among them! Surround them!'"[8]

With the element of surprise lost and Clarke's Confederates starting to surround his troops, Higginson reported that his Black soldiers knelt in the open ground, fixed bayonets, and began shooting as fast as they could load and pull the trigger. Higginson continued, "Some of our soldiers...were actually charging....[The rest] kept together tolerably, however, while our assailants, dividing, rode along on each side through the open pine-barren, firing into our ranks, but mostly over the head of the men.... [B]oth sides had [an] opportunity to do some execution." After an hour, when the fire slackened, Higginson ordered his men to retreat to the boat, being "more than satisfied that we won a victory."[9]

Higginson and Gen. Saxton wrote glowing reports of the great triumph achieved at Township, Florida; Unionists took heart in the success of the USCT. Headlines in the Northern press declared, "A Contraband Expedition Whips the Rebel Chivalry"; "Negro Soldiers, 'Their Gallant Conduct Under Fire'"; and even a headline mocking the familiar Confederates assertion that "Niggers Won't Fight." Higginson put Union losses at one killed and seven wounded, while averring that the Rebels lost twelve, including Lt. John D. Jones. Saxton upped the ante, stating positively that the Secesh lost sixty men in the engagement. Extant records, however, reveal that the graycoats lost exactly one man—Lt. Jones—killed in the fight at Township. A report by a Union soldier stationed at Fernandina casts further doubts on Saxton's version of the story. He wrote: "The Rebels seem to have [gotten] wind of the scheme [operation] and lay low to give the colored solders a rough house....

---

[8] Hartman and Coles, *BRF*, 4:1546–557; "Negro Soldiers, Their Gallant Conduct under Fire," *New York Evening Post*, 10 February 1863, 1; "Niggers Won't Fight," *Council Bluff [IA] Weekly Nonpareil*, 14 February 1863, 2; Schmidt, *Florida's East Coast*, 482; *ORA*, ser. 1, vol. 14, 194–98.

[9] "The War in Florida," *Philadelphia Press*, 11 February 1863, 1.

[T]he colored soldiers were defeated and lost heavily in the scrap." He closed with the detail that one Black soldier had been left in Confederate hands.[10]

At about the same time, USCT registered a series of undisputed successes further west in Nassau County. Aboard the ship *Planter*, Capt. Charles Trowbridge and a thirty-man company of the First South Carolina (USCT) conducted a series of raids on salt-making operations near Amelia Island. They went first to King's Bay, where a previous foray had destroyed a huge number of kettles. They found the site abandoned and began cruising up the Crooked River until they found a major salt-making facility. "After a march of two miles across the marsh, with thirty men, and drawing a boat to enable us cross an intervening creek, we destroyed them [the salt works]," Trowbridge wrote. "There were twenty-two large boilers, two storehouses, a large quantity of salt, two canoes, together with barrels, vats, & c., used [for] manufacturing the salts." He returned to Fernandina without losing a man and inflicted serious damage to the area's Confederate sympathizers.[11]

On the same day that Trowbridge left for Crooked River, Col. Higginson began a second offensive into the Nassau County interior. The commander at Amelia Island needed bricks to make repairs on Fort Clinch, and Higginson used that deficiency as a pretext for another raid. The abolitionist officer and about 250 of his United States Colored Troops boarded the steamer *John Adams*, which Higginson described as "a double ended Ferry boat of large size and powerful engines." The brickyard, which had provided building materials for the original structure, lay thirty-five miles up the St. Marys, where Higginson planned to fill the order. More likely, however, he simply wanted another chance to flex his military muscles and collect additional contrabands for the First South Carolina Volunteers.[12]

The operation began with an early success. The Unionists reached the brickyard without opposition and loaded forty thousand units aboard the

[10] *ORA*, ser. 1, vol. 14, 195–98; "The War in Florida," *Washington [DC] Reporter*, 18 February 1863, 2; "Negro Fighting in Florida," *Amherst [NH] Farmer's Cabinet*, 19 February 1863, 3.

[11] *ORA*, ser. 1, vol. 14, 195–98; "The War in Florida"; "Negro Fighting in Florida." A "barracoon" is a temporary barracks for holding slaves or criminals. Some information suggests that the Albertis, who had been born and raised in the North, were kind, good masters, who even sent one mulatto slave girl to the North to marry her White beloved ("1843: Ernestine [Strong] Alberti to Lydia [Strong] Clapp").

[12] *ORA*, ser. 1, vol. 14, 195–98; "Negro Fighting in Florida"; "The Colored Troops in Action," *Philadelphia Inquirer*, 11 February 1863, 1.

John Adams. Along the trip up the St. Marys they also kept a close eye for a fleet vessel named *Berona*, which was rumored to be hiding up the waterway as it waited for a chance to slip down the river and run the blockade. Additionally, the Black soldiers confiscated a small flock of sheep, five large yellow pine planks, and seven White male citizens to be used as hostages on the return trip to Fernandina. During this time, Sgt. Robert Sutton, who had once been a slave on the Alberti plantation, led a squad to his old homeplace. There he collected the "jewels" from Mrs. Alberti's barracoon. (These devices reportedly included iron collars and wrist and ankle restraints for both sexes. According to Sutton, some of the "jewels" were of Madame Alberti's design and inflicted excruciating pain upon the unlucky person being punished.)[13]

With their mission accomplished, the *John Adams* began the return trip to Amelia Island. They had sailed only a couple of miles downstream when, at a place called Reed's Bluff, Rebel fire suddenly raked the entire vessel. Higginson wrote,

> Suddenly there swept down from the bluff above us…a mingling of shout and roar and rattle as of a tornado let loose; and a storm of bullets came pelting across the sides of the vessel, and through a window, there went up a shrill answering shout from [my] men…. The stream is narrow, swift, and winding, and in many places with high bluffs, which blazed with rifle shots. With our glasses, as we approached these points, we could see mounted men by the hundred galloping through the woods from point to point to await us, though fearful of our shell, they were so daring against musketry that one rebel actually sprang from the shore upon a large boat…where he was shot down by one of my sergeants.[14]

Higginson, of course, minimized his casualties, as leading his men into a bloodbath for a load of bricks and a few pine slats might be deemed imprudent by his superiors. "[N]ot a man of the regiment was killed or wounded," he declared, "though the steamer is covered with bullet marks where our brave Captain Clifton, commander of the vessel, fell dead beside his own pilothouse, shot through the brain by a Minie ball." He did admit that "[t]he secret of our safety was in keeping the regiment below [deck]." Casualty

---

[13] Ibid. Schmidt, *Florida's East Coast*, 486–87. No record of the number of wounded has been located.

[14] Proctor, *FL100*, 1863, 2; Schmidt, *Florida's East Coast*, 487.

reports by the Rebels might be suspect, but a member of the Seventh Connecticut Infantry casts additional doubt on the veracity of Higginson's report. "Defeat number two," he noted in his diary, "was scored with the steamer's Captain and eleven negro soldiers [killed]."[15]

In one of those quirks of timing that occasionally occur in wartime, on 11 February 1863, Gen. Finegan had received an inquiry from Richmond regarding the feasibility of setting a trap for the Yankees on the St. Marys River. The military high command in Virginia "suggested that field guns be placed at strategic positions on both banks of the river, and the Union gunboats be allowed to pass these batteries in ascending the river without drawing fire until within range of the battery highest up the stream. Then, when falling back, the gunboats would be forced to run the gauntlet of the artillery and sharp shooters which would be hidden all along the banks." Col. Higginson, in his postwar histories, wondered why the Rebels did not simply fell trees along a narrow section of the stream, which would have effectively blocked passage up or down the St. Marys. As far as is known, the Nassau County Confederates never used that particular scheme.[16]

JACKSONVILLE

Upon his return to South Carolina, Col. Higginson found orders awaiting him to prepare his First South Carolina Loyal Volunteers for a move to Jacksonville, Florida. They would be gone for at least ten days, and Gen. Saxton explained the importance of the operation. Jacksonville, he wrote, would become "the base of operations for the arming of negroes and securing in that way the possession of [the] entire state of Florida." A Pennsylvania newspaper reported the expectations of most Unionists: "The object of the expedition will be accomplished when the whole state of Florida is brought back into the Union." Higginson and his men left Beaufort on 5 March aboard the gunboats *Boston*, *Burnside*, and *John Adams*.[17]

The Northern journals had loftier expectations for the expedition than the military. The army simply wanted to secure African American recruits to fill the ranks of the newly formed Second South Carolina Loyal Volunteers. Col. James Montgomery, almost always described by pro-Confederate newspapers as "the Kansas Jayhawker," would command the new unit. He had

---

[15] Schafer, *Thunder on the River*, 142.

[16] Ibid.; *ORA*, ser. 1, vol. 26, 423.

[17] Du Pont, *Selection of His Civil War Letters*, 2:326–27.

managed to enlist 125 runaway slaves at Key West for the proposed regiment, but that meager number had depleted the manpower pool at the Union stronghold in South Florida. Gen. Saxton, however, reported,

> I have reliable information that large numbers of able-bodied negroes in that vicinity [St. Johns River Valley] who are watching for an opportunity to join us. The negroes in Florida are more intelligent than any I have seen....They will fight with as much desperation as any people in the world. I have many of these Florida men in the First South Carolina Regiment.[18]

Some Federal loyalists saw the expedition as an opportunity to plunder the state's beleaguered citizenry. Other Union military leaders felt that a permanent occupation would allow the State's "closeted" Federals a chance to express their true loyalty to the Stars and Stripes. The only voice of dissent seemed to be Adm. Du Pont. With keen insight he noted,

> [N]o Union man dare to land in Florida except under the guns of my gunboats, and yet people are sent from home [the Northern states] to tax land which we do not hold. So, to remedy this, they propose to send the black regiment to hold Jacksonville...[I]...told...General Saxton that his regiment could not exist a week in Jacksonville unless I surrounded the place [with gunboats.][19]

Montgomery had a long history as an abolitionist, and like Higginson, he had been a minister and an ardent supporter of John Brown. However, Montgomery's "fervent opposition to slavery did not rest on notions of Negro rights, for he believed blacks to be a savage, inferior people who needed protection from further degradation at the hands of [White] Southerners." Even Higginson seemed to sense a hint of mental instability in the Kansan. The Jayhawker advocated killing all White Southerners, and he would gladly be the instrument of Jehovah's wrath to accomplish that goal. The White Rebels "were to be swept away by the hand of God, like the Jews of old," Montgomery crowed, and he claimed that his men would *not* [emphasis added] be

---

[18] Higginson, *The Complete War Journal and Selected Letters*, 116; Phillips, "Montgomery, James"; Shaw to Annie Shaw, 12 June 1863, Schafer, "Freedom Was as Close as the River," 172; Wilson, "In the Shadow of John Brown," 309.

[19] Schafer, *Thunder on the River*, 143–44; Schmidt, *Florida's East Coast*, 522.

"bound by the rules of regular warfare." Higginson described his fellow abolitionist as "splendid, but impulsive & changeable, never plans far ahead, & goes off at a tangent." Florida's Confederate sympathizers tended to view him as a "lunatic."[20]

The Union vessels again had difficulty in crossing the St. Johns River bar, but the bluecoats arrived at Jacksonville on the morning of 10 March. Higginson's Florida-born men "were wild with delight when they realized their destination." As soon as they debarked, Col. Montgomery, with two companies, quickly spread throughout the city. In the process, they flushed a small body of Rebel cavalry, who immediately fled at top speed. Unionist reporters quickly inflated this encounter into a thrashing administered to the Secesh by the Jayhawker and his Black soldiers: "He [Montgomery] was not long in coming up with a [Rebel] cavalry company, which was attacked and routed with a loss on his part of one killed and two wounded. The Rebels were seen to carry off a dozen or more wounded and dead. A Rebel captain was shot through the head and was carried off."[21]

Lt. Col. Abner McCormick, leading the 2nd Florida Cavalry Regiment, quickly notified Brig. Gen. Finegan of the Federal occupation of Jacksonville. The Irishman, commanding the Department of East Florida, ordered all his units, scattered throughout the district, to converge at Camp Finegan, eight miles west of the River City. He quickly amassed a force of 803 graycoats to confront a Union contingent that soon grew to twenty-five hundred men and three gunboats. (Saxton had reinforced Higginson and Montgomery's Black contingent with White troops of the Sixth Connecticut and Eighth Maine.) Finegan's men came primarily from Maj. T. W. Brevard's Partizan Rangers, and Winston Stephens's, J. J. Dickison's, and W. E. Chambers's companies of the 2nd Florida Cavalry. Finegan also wired Gen. P. G. T. Beauregard for reinforcements. The Creole officer, however, was expecting an enemy attack at either Charleston or Savannah at any moment and had no troops to spare for the military backwater. He informed "Old Barney" that the Florida Confederates would again have to rely on guerrilla warfare and "hit and run" tactics.[22]

---

[20] Schafer, *Thunder on the River*, 150–51; *ORA*, ser. 1, vol. 14, 227–28.

[21] Schafer, *Thunder on the River*, 144–45.

[22] "Our Florida Correspondence," *Augusta [GA] Daily Chronicle*, 26 March 1863, 2; *ORA*, ser. 1, vol. 14, 227.

On the day after the Federals' arrival at the River City, Finegan personally directed a detachment of cavalry into Jacksonville while Maj. Brevard led an infantry force through the suburbs in an effort to spring a trap on the Black soldiers. There are, of course, two distinct versions of what occurred on the morning of 11 March. The "authorized" version, penned by Capt. William Lee Apthorp, a Georgia native serving as a White commander of one of Montgomery's new units, reported that the Rebel attack came as he instructed the raw recruits on how to march and load their weapons. "We waited there for [the enemy's] fire and received it. One man fell severely wounded.... Our men returned the fire coolly as if it was not the first time they ever smelt powder." According to Apthorp, even when his bluecoats ran out of ammunition, they did not lose their nerve or dissolve into a disorderly mob. Though without means to respond to the Confederate fire, "they fell back [in perfect order] to the gunboats...and the repulse begun by us was turned by a few shells and shrapnel into a rout."[23]

The Confederate account states,

> The day after landing, a portion of the enemy's force ventured a mile or so out of town, when they were vigorously charged upon by a small cavalry force, led by Brig. Gen. Finegan in person. The valiant Ethiopians were immediately panic stricken and fled precipitately to town and sought refuge on board their gunboats. Several of the black troops, and one white captain were killed, and great numbers were wounded. Our loss was one killed, Dr. Meredith of Tampa.

The flight of the enemy infantry had been so sudden, Finegan noted, that Brevard's men had not been able to reach the scene before the Federals took cover under the guns of Du Pont's vessels.[24]

The various responses to the accounts of the skirmish were predictable. A Southern journal stated, "The idea of even a handful of our men being backed down by a negro regiment is a slander on [White] freemen." Union officers faced a similar dilemma. They knew that in order for the experiment

---

[23] "The Fight at Jacksonville, Fla.," *Augusta [GA] Daily Chronicle*, 25 March 1863, 1.

[24] *ORA*, ser. 1, vol. 14, 227; "The Negro Regiments in Florida," *Philadelphia Inquirer*, 25 March 1863, 2.

with United States Colored Troops to continue, the Northern politicians and public had to receive glowing accounts of battlefield successes.[25]

Later in the afternoon, Brevard's infantry came back for another try at the invaders. With skirmishers in front, the Rebel foot soldiers advanced into town before encountering a couple of USCT companies near the site of the earlier engagement. The advanced guard pushed forward under fire from Higginson's veterans of the fights along the St. Marys. "Feeling the [enemy] party with skirmishers," Finegan reported, "he [Brevard] then opened on them with his entire command, when they [the USCT companies] broke and fled in confusion, however, [they] returned his fire. Major Brevard then withdrew his command, and although the enemy opened on them with shell[s] from their gunboats, [Brevard] escaped without loss." A Philadelphia newspaper reporter stated, "So far as I have been able to learn, one negro was killed, although many were wounded."[26]

Thereafter, the occupation of Jacksonville fell into a fairly predictable routine. The Federals burned a shantytown along the railroad tracks—originally occupied by free Blacks—to keep the Rebels from using the shacks as cover to slip in close to the Union breastworks. They dug trenches and cut down trees to create defensive perimeters, which the men named Fort Higginson and Fort Montgomery. The Rebels probed the lines a couple of times each day, looking for weaknesses and just to keep the enemy from getting too comfortable.[27]

On one such occasion, Capt. J. J. Dickison blundered into a Yankee trap. A Macon newspaper offered a thrilling account of the incident, writing,

> Major Brevard sent Captain Dickison, with about fifteen of his cavalrymen, to annoy the enemy and induce them to [try to] capture him. He succeeded in drawing them [the Federals] out as far as the brick yard, [where there were] about two hundred of the black scamps, but could get them [to approach] no further. Captain Dickison, being in the rear of his men, was partially cut off, when ten or twelve dusky forms made for him.... He gave his horse

[25] Schafer, *Thunder on the River*, 146–47.

[26] "The Late Fight at Jacksonville," *Macon [GA] Telegraph*, 25 March 1863, 4.

[27] Hillhouse, *Heavy Artillery & Light Infantry*, 54–56; Schafer, *Thunder on the River*, 155–56.

the spur, and leaped down the hill, through the old graveyard, across the marsh, and was soon out of reach.[28]

Though having little to do with irregular warfare, one of the oddest battles of the Civil War occurred near Jacksonville during this time. Finegan knew that as long as the Union gunboats controlled the St. Johns River, the Rebels had little real chance of driving the Yankees from the River City. Capt. Edwin West, a member the First Florida Special Battalion, devised a weapon he hoped would allow the Confederates to compete with the Union gunboats. Mounting a thirty-two-pound cannon on a railroad flatcar, and with Priv. Francis Sollee serving as the gunner, they readied the "Dantean Monster" for battle. On the morning of 24 March, Finegan ordered the railroad gun toward Jacksonville for a trial run. The train engine pushed the weapon to within a mile and a half of town, and the Rebels got off seven shots before the gunboats could respond. The next day, a large Union force of Black and White troops advanced up the tracks toward the Rebels' position. In a perfect example of bringing a popgun to an artillery duel, the bluecoats had mounted a four-inch cannon on a little flatcar to combat the Confederate "Monster." Sollee's first shot hit among the Eighth Maine, killing two and maiming two others. After a very few minutes, the Unionists hurriedly retreated, taking refuge in their entrenchments. The railroad gun and its annoying shelling did little to affect the third occupation of Jacksonville. The Yankees remained safe within the River City, and their gunboats still controlled the St. Johns.[29]

Despite some initial successes, the Federals had accomplished little after three weeks beside making life miserable for the citizens of Jacksonville. Finegan realized that they would eventually move into the St. Johns Valley, and the Irish general worried that the bluecoats might receive reinforcements from St. Augustine. To combat that possible development, he sent Dickison and his men into the borderlands around St. Augustine to prevent any Yankee reinforcements from that post. After that threat failed to materialize, the guerrilla leader would move toward the St. Johns River to do what he could with 110 men.[30]

---

[28] *ORA*, ser. 1, vol. 14, 228.

[29] "From Florida," *Macon [GA] Telegraph* [quoting an article in the *New York Tribune*], 20 April 1863, 3. Montgomery lists the *General Meigs* as the vessel used on the Palatka raid; Thomas T. Russell to Gen. Finegan, *ORA*, ser. 1, vol. 26, 238.

[30] Russell to Finegan letter, *ORA*, ser. 1, vol. 26, 238; *ORA*, ser. 1, vol. 14, 237–39; "The Enemy in Florida," *Richmond [VA] Enquirer*, 13 May 1863, 2;

During his two weeks in Florida, Col. Montgomery had added very few recruits to the Second South Carolina Loyal Volunteers. On 27 March, aboard the *General Meigs,* the bluecoats headed up the St. Johns to collect Black volunteers and supplies. Along the way to Palatka, the Federals dropped off squads of soldiers at various plantations and villages. At the Antonio Baza farm, for example, they "took from him three horses and one cart, [and] all of his poultry, hogs, pots, salt, and everything else they could lay their hands upon." Baza's Black slaves, however, hid out and escaped capture. At Palatka, the Jayhawker wrote that they "received a volley from a company of guerrillas. [However, as soon as they] brought the boat's howitzers to bear upon the town, the rebels scattered in all directions, and soon disappeared in the pine forest.... After obtaining all the cotton at the place, and about thirty negroes for the regiment, Col. Montgomery started his return to Jacksonville." As they sailed north, they picked up their men and the booty the Unionists had pilfered at the various villages and homesteads along the St. Johns. According to Montgomery, it had been a very successful expedition.[31]

Unfortunately, the Union officer's report appears to have been an example of calculated misinformation. Contemporary sources, confirmed by local eyewitnesses to the action, stated,

> When the first man off the boat [who went to] reconnoiter the Palatka area returned to the steamer, the landing began. [Capt.] Dickison waited patiently until there were thirty or forty Yankee targets on the dock and the upper and lower decks were crowded with troops "as thick as they could stand." When the Rebels' firing commenced, the Federals 'immediately retreated to their boat in great confusion...dragging their dead and wounded.

Palatka resident Thomas T. Russell confirmed Dickison's account, adding there was "a great quantity of blood about on the wharf and pieces of bones." Southern estimates of the dead and wounded ranged as high as forty

---

"Depredations on the St. Johns," *Memphis Daily Appeal,* 21 April 1863, 1; B. Michaels, *River Runs North,* 108.

[31] Schafer, *Thunder on the River,* 157–58; "The Evacuation of Jacksonville," *Boston Post,* 13 April 1863, 4.

to fifty, but Dickison guessed a more conservative tally of twenty to thirty Yankee casualties.[32]

On 28 March, vessels from Beaufort arrived at Jacksonville, bringing orders for the Union evacuation of Northeast Florida. Hunter needed all the troops he could muster for yet another attack on Charleston, South Carolina. Higginson and Montgomery expressed disappointment that their troops had received scant opportunity to engage the Rebels in battle. Additionally, the Jayhawker had added very few volunteers to his new regiment, and the residents of the River City, who had aided the bluecoats, felt completely betrayed. Their only options were to leave, abandoning their homes and possessions, or stay and be killed or persecuted by the area's Confederate sympathizers. A furious reporter spoke for most Unionists in the area: "A more fatal order for the place…could not have been made.… Why occupy a place at all, if not prepared to hold it?"[33]

The Federals, however, left a permanent mark on the city before they departed. The Union troops had earlier cut down many of the trees to strengthen their fortifications, and they seemingly set the city afire before leaving, destroying fully two-thirds of the buildings. A *New York Times* writer wrote the following sad account,

> Jacksonville is in ruins. That beautiful city which for so many years [was] the favorite resort for invalids from the North has today been burned to the ground, and what is sad to record, by soldiers of the National [US] army. Scarcely a mansion, a cottage, a negro hut, or a warehouse remains. The long lines of magnificent oaks, green and beautiful, with thickest foliage, the orange groves perfuming the air with their blossoms, the sycamores, the century-old palmettos and [Spanish] bayonet trees, ever tropical in *verdue*, the rose and the

[32] "From Florida," *Macon [GA] Telegraph*, 20 April 1863, 3 (reprinting an article from the *New York Times*).

[33] Schafer, *Thunder on the River*, 160–62; "The Burning of Jacksonville, Fla.," *Portland [ME] Advertiser*, 25 April 1863, 1; "Important from Florida," *Cleveland Plain Dealer*, 9 April 1863, 2; "Review of the Week," *Boston Traveler*, 11 April 1863, 1; "Burning of Jacksonville," *Manchester [NH] Weekly Union*, 21 April 1863, 4; "From Florida," *Augusta [GA] Daily Constitutionalist*, 27 March 1863, 1; "The Enemy in Florida," *Richmond [VA] Enquirer*, 13 May 1863, 2.

> jasmine—all at this season made Jacksonville a little Eden, is now entirely consumed to ashes, by the devouring flames.[34]

Who set the fires has several possible answers—and probably will never be solved. There were many suspects, all based upon the writer's biases. The Southern press, naturally, blamed the Black troops and their White officers. A Virginia newspaper blamed Col. Higginson, who preached to his Black troops that the Catholic and Episcopal houses of worship "were heretic churches and ought to have been destroyed long ago. They were destroyed." A Georgia journal blamed "Montgomery of Kansas notoriety...[who] has commenced applying the torch and committing depredations." Some Northern newspapers concurred, stating, "[T]he black regiment were...implicated in the firing [arson]." The New England press asserted "The Black Brigade had nothing to do with the firing of Jacksonville. That piece of vandalism was the work of white soldiers from...Maine and Connecticut." Col. Rust, of the Eighth Maine, accused "Secessionists" for the conflagration; an Ohio journal asserted the fires were "[i]n retaliation for the conduct of the rebels, Col. Rust...burned the town." Dr. Schafer concluded, "[B]oth white and black soldiers participated" in the destruction. The Confederate company of Winston Stephens arrived almost immediately after the gunboats departed, and, aided by a downpour, saved a part of the town from total destruction.[35]

The lack of success achieved by the three-week incursion into Northeast Florida constituted a major disappointment for the Union high command. The main purpose of the expedition had been to gather enough Negroes to complete Col. Montgomery's regiment and to return Florida to the Union. They had made a single raid, to Palatka, and, in total, the Federals collected about thirty new soldiers. Higginson admitted "he had not yet recruited as many men for the new black regiments as he had anticipated"; modern historian of the USCT, Dr. Dobak remarked on the "small number of Florida slaves who escaped to Union lines" during the third occupation of Jacksonville. For the journalists who had equated success to returning Florida to the Union, the expedition could only have been described as "disastrous."[36]

---

[34] Schafer, *Thunder on the River*, 157; Dobak, *Freedom by the Sword*, 41.

[35] Graham, "Home Front," 39; "Commanders in Divisions," *Charleston [SC] Evening Post*, 11 May 1899, 13.

[36] *ORA*, ser. 1, vol. 16, 739; Evans, ed. (and J. J. Dickison), *Confederate Military History* (Florida), 11:53.

ST. AUGUSTINE

The pro-Confederate bushwhackers stationed around the Ancient City had done a good job of making life miserable for the occupation forces, but they lacked the organization and leadership for a coordinated operation. Though better known for his operations west of the St. Johns River—such as Palatka, Gainesville, and Bridge No. 4—Capt. J. J. Dickison also operated east of the waterway. A historian explained, "The purpose for this force [Dickison's] was to patrol the area west of the St. Johns River watching for Federal incursions and, occasionally, to harass the Federals occupying St. Augustine. Because of the presence of Dickison's troops, the Union forces usually stayed inside the town's defensive perimeter." Some of the few who ventured out into the countryside ended up as the "Swamp Fox's" prisoners.[37]

In late December 1863, Dickison received instructions from Gen. Finegan to take "20 men, with proper non-commissioned officers, and three or four days' rations for the men and forage for his horses, and proceed as secretly and expeditiously as possible across the Saint Johns River in the vicinity of Saint Augustine...to capture...as many men of the enemy or wagons coming into or out of Saint Augustine." Dickison excelled at this type of operation, and a post-war article boasted (with typical Lost Cause hyperbole): "Dickison was everywhere 'and his blows fell like bolts from a clear sky, dealing destruction on every side.'"[38]

The guerrilla leader discovered, from scouts he had posted in the area, that the Federals had gotten in the habit of visiting the Fairbanks house, about a mile and a half from St. Augustine's gates. (The original Fairbanks house, then located five miles west of St. Augustine, had been burned earlier during the conflict.) On 9 January 1863, Lt. Virgil Cate, with seven soldiers and US employees, left the city and proceeded toward the Fairbanks domicile for a "sugar boiling." Dickison and his unit swept down and captured the partygoers without firing a shot. The guerrilla leader and his men re-crossed the St. Johns the following day. A couple of weeks later, a Boston newspaper carried an item, lifted from a Southern journal, which stated, "Eight Yankee prisoners, captured at St. Augustine, by Capt. Dixon's [Dickison's] guerrillas,

[37] "Probable Battle in Kinston, NC," *Boston Herald*, 23 January 1863, 2.

[38] Dobak, *Freedom by the Sword*, 30–34.

arrived this evening at the Oglethorpe Barracks [in Savannah, Georgia]. Among them are a Provost Marshal, sutler, and three merchants."[39]

St. Augustine's pro-Confederate women had earlier received a reprieve from deportation when the vessels had been stopped at Jacksonville and returned home. By 1863, Maj. Gen. David Hunter, a political general who firmly believed in the rights of Black people and using them to fight for their freedom, had command in Northeast Florida. On 9 May 1862, months before the Emancipation Proclamation, Hunter issued General Order No. 11, declaring all African Americans in Florida, Georgia, and South Carolina "forever free." Pres. Lincoln quickly rescinded the directive, but some Union officers in the Deep South had begun covertly organizing groups of contrabands into military companies. Confederate sympathizers in St. Augustine, whether male or female, could expect little sympathy from Hunter.[40]

On 2 February 1863, the steamer *Boston* left the Ancient City "with eighty-three men, women, and children aboard." The deportees could take only a bundle of clothes with them, virtually guaranteeing that their valuables would be "appropriated" as soon as the *Boston* left the dock. The original plan called for the exiles to be delivered to the Confederates at Charleston, South Carolina, but the women became embroiled in a controversy with Lt. Col. Liberty Billings almost before they left the Matanzas River. Billings, a Maine native, Unitarian minister, and avid abolitionist, had been serving as the chaplain for the Fourth New Hampshire Infantry. Even Higginson and Montgomery, commanders of USCT troops, appear to have detested Billings, hinting at his cowardice and immaturity. Billings insisted upon using Black troops to guard the Southern women. The uproar between the two groups became so strident the *Boston*'s captain put into port at Amelia Island. Capt. Frank Clarke's Rebel cavalry soon arrived and served as an escort for "a train of oxcarts…[which departed] in a cold pouring rain, [and] set off for Lake City." From there, the exiled women scattered to Tallahassee, South Georgia, or wherever they could find friends or family who could provide a haven for them.[41]

---

[39] Schmidt, *Florida's East Coast*, 490–94; Graham, "Home Front," 33–34; see also, Asarch, "Liberty Billings."

[40] Graham, "Home Front," 33–34; "The War in Florida," *Philadelphia Inquirer*, 14 February 1863, 4; "From the Coast," *Augusta [GA] Chronicle*, 18 February 1863, 1.

[41] *ORA*, ser. 1, vol. 14, 224–25; Koblas, *Swamp Fox*, 28.

On 9 February 1863, a second, smaller group departed St. Augustine, bound for South Carolina. They consisted of St. Augustine civilians, but the *Boston* also stopped at Fernandina and took aboard an additional number of exiles. Several Northern journals had earlier complained that the mother and aunt of Confederate Lt. Gen. Edmund Kirby Smith had not gone in the first wave of deportees, but they were included in the second group. The exiles noted with alarm a difference in the Black troops. "The negroes were far more insolent than the [White] soldiers, and took great pleasure in insulting the whites; cursing the d—n secesh –'we men;' and laughing [in] their faces. Their insults dare not be resented." Smaller groups would be exiled from North Florida throughout the remainder of the war.[42]

The military situation at St. Augustine had not improved much with the removal of the women. Going a half-mile beyond the city gates put the bluecoats in guerrilla territory, and snipers clung to the line of woods, taking potshots at any unwary pickets. By the first of March, Col. H. S. Putnam noted, "I had reliable information that [the] enemy consisted of a company of about 80 horsemen, commanded by a Captain Dickison, and that his camp was at a place called Fort Peyton, 7 miles southwest of this place."[43]

In an attempt to curtail the annoyance by the Rebel irregulars, Putnam dispatched Lt. Col. Joseph C. Abbott, with 120 men, to capture or disperse the Rebel horsemen. Abbott and his unit crossed the San Sebastian River and advanced toward Fort Peyton. When within three miles of the graycoats' camp, they spotted a couple of Dickison's scouts, who fell back slowly, attempting to lure the Granite State foot soldiers toward the main body, it seemed. Following the accepted military tactics of the time, Abbott sent a squad to the left and right to cover his flanks and then threw forward skirmishers to capture Dickison's outliers. However, by the time they reached the Bartols Masters's house, the guerrillas had vanished. At the domicile the bluecoats questioned Masters, John Manucy, and their families. The Floridians refused to reveal any information except that a couple of wagons, loaded with supplies, had passed thirty minutes before on the Palatka Road.[44]

Abbott deemed "further pursuit useless" and began gathering his men for the return to St. Augustine. The New Hampshire officer concluded his report on the incident, stating,

---

[42] Ibid.

[43] Ibid.

[44] Ibid.

> [T]he men whom I had sent toward the Carrero's house had not reported [back], and on arriving opposite the house I sent a corporal and five men to ascertain the reason. They returned, bringing with them Carrero himself, who stated that the first party had been intercepted by Dickison's (rebel) horsemen, numbering about 80 men, and had been captured.

With no other options remaining, Abbott returned to St. Augustine.[45]

The Granite State officer seemingly never realized that he had missed an opportunity to strike J. J. Dickison's cavalry a crippling blow. Since Finegan had ordered him to operate along east bank of the St. Johns, the Rebel leader had decided to supplement his inadequate supply situation. In his postwar writing, the guerrilla chieftain stated,

> The enemy did not come out in his usual force, nor at the usual time; but six companies, about three hundred and fifty strong...crossed the San Sebastian River...intending to capture our wagon train, which was at the encampment near Moultrie Creek, and cut off the escape of our forces. Lieutenant [William H.] McCardell, with his detachment, held them in check until the train had drawn up in safety.[46]

Incredibly, less than three weeks later, Unionist newspapers from California to Maine carried the report: "From Florida it is learned that there are no armed rebels east of [the] St. Johns River." It was patently false—and almost laughable—but some Federal reporters obviously believed the report. Meanwhile, for several months following the third occupation of Jacksonville, Dickison and his troops "guarded all the country from St. Augustine to [New] Smyrna." When that duty became too arduous, Gen. Finegan would dispatch Capt. W. E. Chambers's unit to reinforce Dickison.[47]

---

[45] M. E. Dickison, *Dickison and His Men*, 47–48; "By Telegraph from Port Royal," *Worcester [MA] National Aegis*, 2 May 1863, 3; "Another Threat of Reprisals," *San Francisco [CA] Bulletin*, 17 March 1862, 3.

[46] *ORA*, ser. 1, vol. 14, 863.

[47] Ibid.; "Department of the South," *Philadelphia Inquirer*, 24 March 1864, 1. Other papers, throughout the North, used this incident to paint all White Southerners as bloodthirsty monsters.

PUTNAM COUNTY (ETONIAH SCRUB)

In late March, another threat to the Confederacy reared its head in the northwest area of Putnam County. On 2 April 1863, Maj. T. W. Brevard issued Special Order No. 18, directing Capt. Chambers to take a dozen men to the "Econiah [Etoniah] Scrub," to check on a possible slave insurrection. Brevard ordered Chambers to "make [a] diligent and careful investigation as to the truth and extent of certain revelations made by the slave Toby, recently arrested in Jacksonville, concerning the alleged conspiracies of the negroes in that section to leave their owners and go to the enemy." Brevard specifically directed Chambers to "act coolly and temperately, and not to use unnecessary harshness."[48]

As far as is known, Chambers, who was seriously ill and would soon resign his commission, left no report of his expedition along the Putnam and Clay County lines. Almost a year later, however, his unit's actions would become a cause célèbre in Northern newspapers. In 1864, the *Philadelphia Inquirer* reported,

> [T]wo negroes, both slaves…ran away from their masters, crossed the St. Johns, and were making their way to [the Federal garrison at] St. Augustine. These slaves were pursued and captured by CAPTAIN CHAMBERS['] Company, East Florida Cavalry. Chambers's troop publically hanged the black men and buried them. Upon discovering that one of the unfortunate black men still lived, they dug him up and finished him off with their gun butts.

Chambers—if he was even present—had obviously forgotten Brevard's admonition to act "coolly and temperately, and not use unnecessary harshness."[49]

With so many White men away in service of the Confederacy, the guerrillas and Home Guard units took the place of the prewar slave patrols. "[T]he heightened demands of wartime security," Sutherland writes, "required even keener vigilance and more formidable deterrents." The presence of White Tories increased the difficulty of their task. The local Unionists often encouraged the slaves to seek freedom, or, at least, gave aid to those who fled their

---

[48] Sutherland, *Savage Conflict*, 46–47; "The Enemy in Richmond," *Richmond [VA] Enquirer*, 13 May 1863, 4.

[49] Sutherland, *Savage Conflict*, 46–47; "The Enemy in Richmond," *Richmond [VA] Enquirer*, 13 May 1863, 4.

masters. "Thus, irregulars policed the slave population as carefully as they did white Unionists," Sutherland continues, "and roving bands of guerrillas...discouraged slave rebellions. They retaliated quickly and forcefully to signs of unrest, and [their] stealthy presence endangered slaves who wandered too far from home."[50]

CEDAR KEYS AND LEVY COUNTY

By the early months of 1863, the Federal Navy had made considerable progress in halting blockade-runners from slipping in and out of major Southern ports such as Charleston, Savannah, and Mobile. Still, "many inlets, lagoons, rivers, and bays of the Florida coast were ideal for the purpose of the daring blockade runners who...delivered cotton [and turpentine] in exchange for arms and ammunition," as well as luxury goods that could be sold at exorbitant prices. The Federals recognized the existence of these gaps in their cordon and set about to choke off the smaller, more remote waterways such as the Crystal, Withlacoochee, Waccasassa, and Suwannee Rivers.[51]

Since the 4 October 1862 ambush of the Union sailors by salt workers, or Rebel guerrillas, at Cedar Keys, the blockading fleet had paid particular attention to the Levy County coastline. Not content to patrol the Gulf islands, and those in Waccasassa Bay, the Union sailors began making forays into the interior. Raiding the swampy hinterlands should have been a relatively risk free endeavor for the bluecoats. According to a Georgia newspaper, "Out of one hundred and ninety voters [in 1860] in Levy County, she has one hundred and forty men in the field [military service]." Capt. Edwin West's small company, stationed primarily at Fowler's Bluff (along the lower reaches of the Suwannee River), could not easily or quickly respond to Union raids, leaving primarily Lutterloh's tiny band to guard the areas adjacent to Cedar Keys.[52]

In an attempt to encourage the officers and men of the blockading squadron to do the best possible job, members of the US fleet received a share of the money realized by the sale of captured Rebel ships and their cargo.

---

[50] *ORA*, ser. 1, vol. 23, pt. 2, 172–73; "Well Done, Florida," *Macon [GA] Telegraph*, 6 February 1862, 2; Starkey, "1st Florida Special Cavalry."

[51] Strickland, "Blockade Runners," 86–87; Wise, *Lifeline of the Confederacy*, 110–12.

[52] *ORN*, ser. 1, vol. 17, 460, 512; "A Blockade Runner—Capture of Cotton," *Washington [DC] Daily National Republican*, 20 August 1863, 2.

> The Federal policy was not to sink or destroy blockade runners, if they could be captured without resistance...each man on a Union ship had a share in the captured Rebel prize, which was turned over to the Union Navy or sold by the Admiralty Court in Key West. The captured ship *Memphis* [for example], paid $510,914.07, and the *Banshee*...paid $104,948.48 to their [the captor's] lucky crews.

The sale of the ship, however, did not always mean the end of its service as a blockade-runner. A contemporary observer wryly noted, "Many mysterious transactions occurred at Key West," and most of the vessels "inexplicably" wound up in the hands of Southern entrepreneurs. The captains and men who successfully ran the blockade received more wages for a single trip than most would garner in ten years of hard work. So, the captains and crews of both sides had plenty of incentives for doing a good job.[53]

During the spring, the *Fort Henry* and her sister ships kept busy scouring the Levy County coast for blockade-runners. On 22 May, they captured the sloop *Isabella*, loaded with twenty-nine hundred pounds of shelled corn; eight days later they seized a "copper-fastened" boat laded with 56 bales of cotton. On 17 June, Acting Lt. E. Y. McCauley sent his men, in launches, up the Waccasassa River. They captured twenty-two bales of cotton and also brought out eight slaves.[54]

It took the Levy County Rebels less than a month of inland raids before they decided to use Union greed as their ally. An official report, from the *Fort Henry*, noted, "In passing the Waccasassa, a large amount of loose cotton floating downstream attracted the attention, and the possibility of making a good cotton capture, induced Boatswain Mate [Frank] Gillespie to pull up the river." When Gillespie and his volunteers reached a narrow section of the stream, area residents, estimated to number fifty to sixty men, opened a deadly fire on the Federal boats. Caught in a trap, two brave sailors, identified only as "Doran and Bishop," sprang forward to bring the boat's swivel guns into action, but Doran died instantly, and Bishop lingered an hour before expiring. Other men finally brought the guns into action, and Gillespie and his men beat a hasty retreat. The wounded, if any, were not enumerated, nor did the journal identify the guerrillas who sprang the ambush.[55]

---

[53] *ORN*, ser. 1, vol. 17, 513.

[54] "Overview...Waccasassa, Wekiva, and Otter Creek Paddling"; "From Florida," *Augusta [GA] Chronicle*, 31 July 1863, 1.

[55] Ibid.

A similar incident occurred along the Wekiva River during the second week of July. (The source of that waterway is a spring in Gulf Hammock, and the Wekiva meanders for eight or ten miles through the Levy County piney woods and marshes before emptying into the Waccasassa a few miles from the Gulf.) A number of plantations and farms dotted the area along the northern section of the stream, and that region had previously received little attention from the Unionists.[56]

According to a Georgia newspaper,

> Two Yankee barges, each containing fifteen to twenty men, went up Weekeeva [Wekiva] Creek, in Levy County, Florida, some ten miles to the plantation of Col. Thompson, and fired on a gang of negroes who were working in the fields. Not one of the negroes were killed but the enemy succeeded in capturing three of them, whom they carried off.... They then visited several other plantations in the vicinity, and at every house they destroyed furniture and every thing of value they could not carry away. They stole six bales of Sea Island cotton...and about sixteen [additional] bales from Messr. Gonzales.[57]

Apparently, some of the captured slaves told the raiders that a large quantity of cotton had been secreted in the area, and the Federals determined to seize it for the prize money. "The party returned the next day," the journal continued,

> intending to carry off several hundred bales of cotton from the plantations in the vicinity. In the meantime the people of the settlement had organized...themselves for defense, and concealing themselves on the bank of the creek, [and] when the Yankee barges made their appearance, opened a rapid, effective fire on them. Seven of the enemy were killed and a number wounded, [whereupon] the barges made a precipitate retreat.[58]

---

[56] Ibid.; Proctor, *FL100*, 20 July 1863.

[57] Ibid. The two units—Mooty's and Gary's—hailed from nearby Marion and Alachua Counties. The following year they would be dismounted and assigned to the 9th Florida Infantry Regiment, Army of Northern Virginia.

[58] "Interesting from Key West," *New York Herald*, 20 January 1863, 2; Grismer, *Tampa*, 142.

The participants sent a report of the engagement to Gen. Finegan, who immediately dispatched the home guard companies of Captains A. P. Mooty and S. M. G. Gary "to protect the plantations from further molestation, and give the enemy a warm reception, should they again venture into that vicinity." The two companies stayed in Levy County only briefly, but the actions of locals, and presence of Confederate regulars, kept the bluecoats away from that locale for a while. The Federals had earlier boasted they would next raid the Otter Creek area, but no record exists of a foray into that region.[59]

TAMPA

The situation at Tampa had changed very little during the first months of 1863. The Federal Navy harassed the Confederates, and Capt. Pearson replied in kind. The *New York Herald* reported in mid-January, "The *Beauregard* has been of late, throwing a few thirty pound Parrott shells into Tampa, destroying some houses, but affairs of this kind accomplished nothing except frightening women and children." It represented the type of casual violence that raised no eyebrows in the North or South, but which caused some Tampa families to move farther from the gunboats on the bay. A local historian reported, "Tampa had become almost a ghost town. Nearly everyone who was financially able had moved to the interior. Many settled in the rural communities of Alafia, Keystone, and Cork. To the latter place...all the county records were taken for safekeeping."[60]

Short-tempered, and totally dedicated to the Confederate cause, Capt. John W. Pearson had been on the receiving end of the Union "outrages" long enough. He had seen Union gunboats enter Tampa Bay flying the Stars and Bars to capture blockade-runners, and he had seen attacks on civilian property, such as the Abel Miranda homestead. As a result, he waited for a chance to strike the hated Yankees a blow that would both hurt and embarrass them.[61]

A Union account described Pearson's trickery.

> The [US gunboat] *Pursuit* had been at Tampa Bay Florida, and while there [at Gadsden Point], it appears three rebels, disguised as [Black] women were seen on shore waving a white flag to attract

[59] Waters, "Tampa's Forgotten Defenders," 13.

[60] Ibid.; "Department of the South," *Philadelphia Press*, 14 April 1863, 2.

[61] Waters, "Tampa's Forgotten Defenders," 13; C. Brown Jr., *In the Midst*, 85.

> attention. A boat, with an officer and ten men, was immediately sent to their assistance from the bark [*Pursuit*]; but when they landed, a body of fifty or sixty rebels rose from the bushes and fired at the boat, wounding five of our men. The crew jumped overboard, and using the boat as a barricade [protection], pulled off the shore, at the same time firing at the enemy. One of the disguised rebels was shot. The Union officer in command was wounded in the arm which will probably be amputated.

This "outrage" only served to fan the flames of hatred, and both sides searched for a way to get the upper hand in the Tampa Bay region.[62]

CATTLE COUNTRY (SOUTH FLORIDA)

The situation in South Florida had deteriorated into a civil war, pitting neighbor against neighbor—and sometimes brother against brother, or father against son. The Confederate government—desperate to fill the ranks of state's soldiers lost in the gory combat in Tennessee and Virginia—badgered the states to press into service almost anyone who could walk and shoot a gun. Capt. Pearson acted with his usual fervor. Historian Canter Brown Jr. explained, "As time had passed, divisions between local Unionists and secessionists had deteriorated to the point that a Regulator organization had been revived to inflict vigilante penalties on those who remained loyal to the United States." Pearson made enforcement of the conscription act a "top priority."[63]

William McCullough, who had fought with the US Army during the Second Seminole War and remained a Federal loyalist, described the methods used by the Southern sympathizers. He wrote,

> The Regulators got after me for my fidelity to the [US] flag. They threatened my friends with death by hanging, and confiscation, or with the threat that they could not be allowed to live in the country. This caused nearly all my adherents to leave me, and some of them even became my persecutors and betrayers, implicating several of my best friends. When I saw that I could do nothing more in the cause of the old flag, and that the regulators were determined[,] with Capt. Pearson's Conscript Officers[,] to take me and

---

[62] C. Brown Jr., *In the Midst*, 86; McCullough, "My National Troubles," 60.
[63] R. Taylor, *Rebel Storehouse*, 100; R. Taylor, "Cow Cavalry," 196.

> my friends dead or alive, I made up my mind to settle my business and leave for Indian territory.

McCullough eventually became an officer in the Second Florida Cavalry (US).[64]

During the initial months of 1863, the South Florida region became extremely important to the Southern Confederacy. The 4 July 1863 surrender of Vicksburg, Mississippi, meant that the Rebel armies could no longer rely on livestock transported from Texas and the Trans-Mississippi region to feed their soldiers. (The loss of Vicksburg was more symbolic than reality—the flow of beef from the Western Confederacy had long past become a trickle rather than a steady stream.)

> Federal patrols ranged up and down the [Mississippi] river and stopped any attempt to move supplies of any kind across [the waterway]. The Confederate leadership now found itself depending on Florida beef to subsist its troops as it never had before. The loss of the Mississippi coupled with the occupation of most of Tennessee by the Yankees, tightened the screws on the South and made it increasingly difficult for the Commissary Department to provide any sort of rations.[65]

The huge herds of half-wild cattle that roamed the prairies of South Florida came from the Andalusian cattle brought to the New World by the Spanish *conquistadors*, settlers, and monks. By 1860, estimates of the size of the herds in the southernmost state numbered at 658,609 though that appears to have been a conservative guess. The herds in northern Florida near the railheads and main centers of population, had already been exhausted by 1863, and the best hope of obtaining food for the Confederate armies lay in the state's southern reaches. Wealthy cattlemen and brokers, such as Jacob Summerlin and James McKay, had earlier established a market for South Florida beef in Cuba: and they had even "smuggle[d] a few Florida beeves to the West Indies…between the summer of 1862 and October 1863."[66]

---

[64] Akerman, *Florida Cowman*, 82–85; Akerman and Akerman, *Jacob Summerlin*, 56.

[65] "From Florida," *New York Evening Post*, 18 June 1863, 3.

[66] *ORN*, ser. 1, vol. 17, 370–71; Proctor, *FL100*, 2 March 1863.

It did not take long for the Northern generals and politicians to recognize the military importance of South Florida's vast herds. An unidentified correspondent noted,

> Florida is now the greatest resource of the rebel army for beef, and since the communications with Texas were partly stopped by the operations on the Mississippi, thousands of head have been weekly gathered here and transported to the west and north. If the St. Johns River had been held, and Black troops alone been continued at Jacksonville, that place would be at this moment the most important base of effective operations against the enemy, and would cripple them in supplies, as well as important territorial advantages.

The task of maintaining an avenue to drive cattle to the Confederate supply centers would become a job for the guerrilla companies of Florida.[67]

NEW SMYRNA

Since the March 1862 engagement between the Confederates and sailors from the *Henry Andrews* and *Penguin*, New Smyrna and Mosquito Inlet had settled into a kind of wartime normalcy. The US blockaders continued to check the "keyhole" port, and blockade-runners still utilized the tiny haven in preparation for the short trip to and from the Bahamas. In late February 1863, Acting Rear Admiral Theodorous Bailey, the new commander of the East Coast Blockading Squadron, directed Acting Master's Mate Jeff A. Slamm, of the USS *Sagamore*, to track down and destroy the fast blockade-runner, *Florence Nightingale*. Bailey had apparently received information of the sleek vessel's location and valuable cargo.[68]

On the last day of February 1863, the *Sagamore* located her prey at anchor in Indian River, near New Smyrna. The various Union accounts of the ensuing engagement reported the Federals achieved a solid victory. The *New York Tribune* reported: "[A] brisk engagement of twenty minutes occurred resulting in the destruction of the vessel, and the loss on our side of one killed

---

[67] Ibid.; "The War News," *Philadelphia Inquirer*, 19 March 1863, 2; "News of the Day," *New York Tribune*, 19 March 1863, 4; "Eastern Gulf Squadron," *Philadelphia Press*, 19 March 1863, 2.

[68] "An Affair in the Harbor of New Smyrna, Fla.," *Macon [GA] Telegraph*, 18 March 1863, 1; "The Affair at New Smyrna," *Charleston [SC] Courier*, 21 March 1863, 4.

and five wounded." A Pennsylvania journal enthused, "The United States steamer *Sagamore* has recently made a successful trip along the Florida coast...during which the officers and crew succeeded not only in destroying a vessel loaded with cotton...but also clearing the Rebels out of Indian River, a stream of one hundred miles in length, in the eastern part [of Florida] flowing parallel to the coast. The trade between this section and Nassau has for many months been of great pecuniary advantage to the parties engaged." Another newspaper stated that the Union raids "have been of efficient service in clearing the rebels from Indian River, and in breaking up their connection to the lawless hordes of Nassau." The results of Slamm's expedition consisted of the capture of "a sloop and a boat with thirty-five bales of sea-island cotton."[69]

Not surprisingly, Confederate accounts told a very different story. One stated,

> On Monday morning, about 7 o'clock, four boats were visible within a quarter mile of the schooner [*Florence Nightingale*] before any one saw them from the vessel or shore...Captain Martin...had made preparations to burn her, and when the first boat was within two hundred [feet] he applied the torch and left her. As the [Union] boats neared, a discharge of ten or twelve rifles and shotguns poured into them, and many a Yankee fell in his tracks. They came up cheering, and in good humor but when the first volley was fired it seemed to change their spirits greatly. They then discharged their boat howitzers two or three times at us, and then turned and ingloriously skeedaddled.

A second article even reported that Martin and his men put out the fire and limped off toward Nassau. The first article concluded, "This is the second time they [the Federals] have been repulsed at this place, and if the people continue to show the same spirit.... New Smyrna will soon be too hot a place for the [Northern] miscreants, unless they come in overwhelming force."[70]

Obviously, neither side had a patent on the truth, but what happened next lends some credence to the Southern reports. In July 1863, a quiet Sunday morning, a pair of Union vessels—the *Oleander* and *Beauregard*—entered the channel at Mosquito Inlet, capturing one sloop loaded with cotton and another empty ship. The Union vessels then began a bombardment of the

---

[69] *ORN*, ser. 1, vol. 17, 529–30; Sweett, *New Smyrna*, n.p.

[70] Ibid.

tiny hamlet, destroying "all the building occupied by troops." According to one account, the Union landing party "was fired upon [by] stragglers concealed in the bushes, but if the bushwhackers thought they could discourage the Union sailors, their weak volley only increased the bluecoats' anger." The first hint of trouble for the civilians came when an artillery shell crashed into the Sheldon Hotel, "slicing the top off the piano and sending splinters flying. The astonished and frightened people gathered their small children and the things they could quickly get together and fled to the shelter of the woods." Harassed by swarms of mosquitoes, the Shelton family lit a smudge pot, but the light from the smoker attracted the attention of the gunners on the ships. As a result, the Sheldons retreated further into the forest. By dawn, the town of New Smyrna had been completely destroyed, but the Sheldons, Longs, and their neighbors would, in time, rebuild their homes.[71]

[71] Ibid.

Chapter 5

"To Keep the Rebel Armies Fed and Fighting"

(August–December 1863)

SITUATION

During the last months of 1863, the guerrilla war shifted from the more populous area in Northeastern Florida to the sparsely settled rangeland and swamps of South Florida. Demands for cattle and foodstuffs increased dramatically during that time. One such appeal, from Maj. J. F. Cummings, Gen. Braxton Bragg's commissary chief, explained the situation bluntly. "We are now dependent upon your state for beef," he wrote. "The future of the army depends on how well it is fed, and this in turn depends upon our ability to secure food from Florida." The state that the Confederacy had abandoned in early 1862 and ignored—except to syphon off its troops for the armies in Virginia and Tennessee—would now play a vital role in preserving the Confederate armies. As Dr. Robert Taylor observed, "Florida no longer remained a secondary source [for foodstuffs]; it now moved to the forefront in plans to keep the rebel armies fed and fighting."[1]

With the desperate need for Florida beef to feed the armies, South Florida cattle barons such as Jacob Summerlin, Francis A. Hendry, and Louis Lanier redoubled their efforts to collect and drive cattle to the railheads in South Georgia. The drovers faced immense problems in moving the cattle out of state. In fall 1863, Maj. Pleasant White, a loyal Confederate, and described as "an indefatigable officer," tried to acquaint the Rebel commissary department with the difficulties the South Florida department faced:

> These problems included the scattered nature of the herds in the fall; a continuing shortage of experienced herders...the lack of adequate grazing lands for animals being driven north; speculators who paid more for hides and tallow than government agents could pay for the entire steer; and continuing transportation difficulties exacerbated by the lack of a railroad connection between Georgia and Florida.

---

[1] Proctor, *FL100*, 5 October 1863; R. Taylor, *Rebel Storehouse*, 98.

Despite those hardships, "from November 20, 1863 to December 16, 1863, approximately thirty-five hundred South Florida cattle" reached the Confederate armies in Tennessee, South Carolina, and Virginia.[2]

Just driving the herds was a difficult endeavor. The cowmen had to traverse a wilderness region inhabited by bears, panthers, wolves, Unionists, deserters, and other predators. A member of Hendry's "Cow Cavalry" unit left a vivid description of one such drive. "A detail of six men," he wrote,

> under the command of Mr. James F. McMullen, was ordered to the cattle pens at Fort Meade to take charge of 365 beef cattle bound for Savannah [Georgia]. Early one morning the drive was commenced. The course was a northerly one, and in line with Orange Lake in Marion County [which is surrounded by a vast savanna called Paynes Prairie]. The cattle were driven along at a "grazing rate" of speed, usually averaging around eight and one half miles a day. At night, if we were fortunate, we reached a cattle pen, and here the beeves would be corralled until morning. With the coming of day the drive started again, the cattle grazing over the countryside at a pace set by the animals themselves....From Orange Lake we continued to the state line, crossing the St. Mary's River near the ferry at Traiders [Trader's] Hill. All was uneventful until the herd came to fording the Altamaha River some miles north of Brunswick, Georgia...[There, two cows were drowned crossing the river, and the hungry locals dressed them out in less than an hour, "leaving not one horn, hoof, or tail."] From there we drove the cattle to Savannah, where they were delivered—Three hundred and sixty-two head out of the original herd of 365 had made it.[3]

Florida beef continued to move north to feed the Confederate armies until late 1864. Sherman's "March to the Sea," and the capture of Savannah, "disrupted communications and made it unsafe to move any more beef northward." However, "[c]attle moved again in the spring [1865] and continued until the end of the war." The exact number of Florida scrub cows that

---

[2] Coles, "Cattle Wars," 95, 99; Akerman and Akerman, *Jacob Summerlin*, 55.

[3] Ivey, "Accidental Pioneer," 55–56.

reached the Rebel armies will likely never be known, but best guesses range from seventy-five thousand to one hundred thousand.[4]

SOUTH FLORIDA—TAMPA

The troubled tenure of Capt. J. W. Pearson, as commander at Tampa, came to an end in September 1863. During his time in Hillsborough County, he had diligently served the Confederate cause. He had been a thorn in the side of both the Federal military and local Unionists. Pearson's strict enforcement of Confederate conscript laws and confrontational attitude seemingly alienated all but the most hardcore Rebels. Though no documentation has been found, it appears likely that his "gung-ho" attitude led to his reassignment. Pearson took great pride that, during his tenure, Tampa remained the only major port city in East and South Florida still in Confederate hands.[5]

Word that Pearson had been reassigned reached Adm. Theodorous Bailey, heading the East Gulf Blockading Squadron at Key West, very quickly. Union sympathizers also noted that James McKay had resumed his lucrative business of running the blockade and planned a make a run to Cuba in the very near future. Bailey quickly decided that he could kill two birds with one stone. He would send a large force to Tampa to burn McKay's vessels and to destroy a "large salt works factory," also owned by the Scottish businessmen. In operation since early in the war, the works consisted of large boilers, giant kettles, vats, and barrels. Both sides agreed that the loss of that facility would be "a devastating blow to the people of Tampa." Bailey also hoped, if he moved quickly enough, the Bay City could be taken before Pearson's replacements arrived.[6]

Bailey and Lt. Com. A. A. Semmes, commanding the gunboat *Tahoma*, jointly devised the Federal battle plan. Two boats, the *Tahoma* and *Adela*, would advance into Tampa Bay, but when they reached Gadsden Point, the Yankees would put ashore a one hundred-man landing party. It would take the raiders several hours to traverse the ten miles from Gadsden Point to the Hillsborough River just west of Tampa and Fort Brooke. (Two men who had

---

[4] R. Taylor, "Rebel Beef," 30–31.

[5] Waters, "Tampa's Forgotten Defenders," 14–16. Pearson and the Oklawaha Rangers fought first at Olustee then became part of the 9th Florida Infantry Regiment in the Army of Northern Virginia. Pearson received a mortal wound in August 1864 leading a charge at the Battle of Globe Tavern. He is buried in Savannah, Georgia.

[6] Ibid.; Zerfas, "The Hillsboro River Raid and the Battle of Ballast Point."

formerly lived in the area, Henry A. Crane and James H. Thompson, would guide the Federal raiders on the trip overland to Tampa.) To keep the garrison distracted, the *Tahoma* and *Adela* would bombard the town with an array of artillery that might frighten even seasoned veterans. The *Tahoma* "carried an 11-inch Dahlgren (200-pounder) pivot gun, two twenty-pounder Parrotts, and two 24-pounder smoothbore guns." The *Adela* also boasted a pair of large-bore rifled howitzers.[7]

On 15 October 1863, the Union operation began. Luckily for the Confederates, Capt. John Westcott, and his Swamp Marines, arrived at Tampa just in time. Westcott's men had been operating at various East Florida locales, but the officers and politicians at Tallahassee seemingly hoped South Florida residents would welcome Westcott as a change from the fiery Pearson. The 56-year-old officer, a New Jersey native, had briefly attended West Point before moving to Florida. Westcott was a medical doctor, surveyor, veteran of service in the Seminole Wars, and owner of a sawmill in Madison County. In addition, Westcott's active mind seemed to always be dabbling with various mechanical devices. Perhaps more importantly, the people of South Florida were familiar with Westcott. In an 1858 congressional election, Westcott "[had received] much support from the southern counties." His chief backing came from Democrats and ex-"Know-Nothings" and was based on his platform calling for "cheap land, cheap money, and payment of Indian War volunteers." Additionally, many of his troops "were local men, described as 'old Indian hunters, who [had] bushwhacked with the Indians but a few years ago, and beat them at their own game.'"[8]

Lt. Com. Semmes's vessels left Key West on 12 October 1863, and the Yankee sailors might have found Tampa largely undefended, but a series of misadventures delayed their operation for two days. Though they reached Tampa Bay on 13 October, the *Tahoma* ran aground three times; the *Adela*, while towing her sister vessel off the bar, experienced engine failure. Despite these problems, on the morning of 16 October, the Federal ships sailed to within two thousand yards of the city and began their bombardment.[9]

---

[7] Ibid.

[8] Waters, "Tampa's Forgotten Defenders," 14–16; Zerfas, "Hillsborough River Raid"; Dillon Jr., "Civil War in South Florida," 26, 210; Knetsch, "John Westcott and the Coming of the Third Seminole War," 5–16.

[9] Zerfas, "Hillsboro River Raid"; Grismer, *Tampa*, 145.

The huge guns created havoc in the town, and the soldiers and civilians took cover wherever it could be found. Many townspeople simply fled into the nearby woods. During the fireworks, a "40-pound [shell] fragment entered the home and flew across the dining room table where [Unionist leader Henry A.] Crane's daughters were sitting. Luckily, no civilian injuries were reported." According to local folklore, during a halt in the cannonading, Crane sent a note demanding the Confederates hand over his son, a member of Westcott's company, "so that the damned rebel could be hanged from the smokestack of the *Tahoma.*" In reply, young Crane publicly declared that he would gladly hang his father "from the highest oak in the courthouse square," if he ever laid his hands on him. Near nightfall, the two boats moved away from the town and under cover of darkness dropped off the one hundred-man attack force at Gadsden Point.[10]

The night trip to the Hillsborough River, where workmen were busily preparing McKay's two blockade-runners, proved to be an arduous journey. Adding to the difficulty of the operation, James H. Thompson, one of the guides, fell ill and had to be carried on a litter virtually the entire way to the Bay City. The bluecoats arrived before dawn and managed to get a couple of hours of rest before beginning their assault. Shortly after daylight the Federals located McKay's ships. "Acting Ensign Joseph P. Randall, and a 'suitable number of men,' boarded and rowed out to the two vessels," Grismer wrote, "where they captured five of the seven crewmen, two having escaped to [the opposite] shore." The Federal raiders applied the torch to both of McKay's vessels—the *Scottish Chief,* laden with 156 bales of cotton, and the *Kate Dale.*[11]

According to local folklore, Joseph Robles, a man of Spanish descent and standing barely five feet tall, saved McKay's salt works. Using a combination of bluff and a ten-gauge shotgun, Robles drove off the party sent to destroy the warehouses and boilers. He reportedly captured several of the raiders.[12]

The two men aboard McKay's ships who had escaped the raiding party immediately spread the news of the Yankee incursion. Reacting quickly, Westcott led his Swamp Marines, and a cow cavalry unit of about forty men, in pursuit of the enemy. In addition to their side arms, Westcott's men

---

[10] Ibid.

[11] Ibid.

[12] Ibid.

dragged along one of Pearson's small-bore cannons. The artillery piece propelled round balls that scattered, like buckshot from a shotgun, when fired. Union Surgeon William Gale reported he "took a four ounce lead ball from another of the wounded that had apparently come from the homemade...Rebel field piece."[13]

Thus far, during the raid, the Federals had maintained the upper hand, but their situation rapidly deteriorated on the retreat. "[J]ust as the bluecoats reached the shoreline opposite the *Tahoma*, Westcott's troops exploded from the brush...[firing and screeching the Rebel Yell]. A running battle ensued with the Yankees scrambling through the surf." As the frightened Federal raiders floundered in the water, intent on gaining the safety of their vessels, some of the Rebel horsemen rode into the water in a vain attempt to cut them off from the ships. According to an eyewitness account, by a Southern soldier, panic gripped the retreating Yankees: "We captured sixty-two Enfield rifle[s], and a few swords and revolvers. Hard bread and cheese, pork & c. was scattered all over the palmettos as they ran. They crawled in the water to their boats, about four hundred yards [away] to avoid the [musket and cannon] balls. Our men [along the shoreline] poured in a continuous fire." The commander of the *Adela* began firing artillery into the woods, but it did little to deter the Rebel fire. In an effort to find shelter from the hail of bullets raining around them, some of the raiders flipped over the launches, but the wooden barriers offered scant protection from the powerful rifles. After what must have seemed like an eternity, most of the Yankee raiders reached the safety of the gunboats.[14]

Both sides claimed victory. The Rebels and Yankees each suffered about twenty casualties, but the damage to Tampa had been extensive. Westcott proved that he and his men would fight and could not be bluffed into retreating by a show of force. Historian Rodney Dillon wrote, "The Confederates had managed to provide a strong, effective defense and inflict casualties on the Federals...but the Union expedition had accomplished its goal—the destruction of the blockade runners."[15]

---

[13] Ibid.; Pizzo, *Tampa Town*, 69.

[14] Zerfas, "Hillsborough River Raid,"; Dillon Jr., "Civil War in South Florida," 213–15; Waters, "Tampa's Forgotten Defenders," 14–16.

[15] Waters, "Tampa's Forgotten Defenders," 15; Dillon Jr., "Civil War in South Florida," 215–16; Zerfas, "Hillsborough River Raid"; "Late Affair at Tampa, Fla.,"

The war in South Florida soon settled into a strange equilibrium. The Rebels continued collecting and driving cattle to Georgia, and the Federals continued their blockade of the Eastern Gulf of Mexico. On 24 December, the Union Navy delivered its Christmas gift to Tampa in the form of a two-hour bombardment. Fearing that this might be a diversion to cover another raid, Westcott quickly scrambled his men to their duty stations, but the gunboats soon returned to Key West.[16]

SOUTH FLORIDA—CATTLE COUNTRY

A series of minor skirmishes occurred in December 1863 that served as a precursor of future martial actions in the South Florida region. By fall 1863, the Union military and political leadership had recognized the importance of the South Florida cattle country and set in motion plans to halt the shipment of Rebel beef. Brig. Gen. Daniel Woodbury, commanding the District of Key West and Tortugas, began by establishing a military presence at Fort Myers. The old Seminole War bastion "became a major source and symbol of Union power in Southern Florida. The fort provided the bluecoats a springboard to raid the countryside, and also served as a safe haven for the area's Unionists, Confederate deserters, and, runaway slaves."[17]

During this period, the Federals also began enlisting White Southerners to form a regiment. The initial idea for the unit came from Enoch Daniels, a Southern Unionist who had led a volunteer company during the Third Seminole War. The new outfit, numbering fewer than fifty men, took the name "Florida Rangers," but that unit would later form the nucleus for the United States Second Florida Cavalry. Daniel "believed that if the refugees could be mustered into service, armed, supplied, and backed by approximately one hundred northern troops, they could encourage Unionism in southwest Florida and gather a force sufficient to break up cattle drives in the area."[18]

The East Gulf Blockading Squadron provided able assistance for Capt. Daniel's unit during its formational period. They began by establishing "a

---

Philadelphia Inquirer, 23 November 1863, 1; and various reports by US Naval officers, *ORN*, ser. 1, vol. 16, 570–77.

[16] *ORA*, ser. 1, vol. 28, pt. 1, 751.

[17] Dillon Jr., "Civil War in South Florida," 223–36; Buker, *Blockaders, Refugees, & Contrabands*, 116–24.

[18] Buker, *Blockaders, Refugees, & Contrabands*, 116–24; Dillon Jr., "Little Affair," 316.

refugee community on Useppa Island in Charlotte Harbor." (During the early days of its existence, the "Florida Rangers" often joined forces with the 47th Pennsylvania Regiment, which was also stationed in the Charlotte Harbor area.)[19]

Anxious to show what his new unit could do, Daniel began with a raid along the Myakka River on Christmas Eve 1863. His force consisted of twenty Florida Rangers; twenty-five men from the Forty-seventh Pennsylvania; a Squadron ship, *Rosalie*; and a naval detachment, which would provide support during the foray. Daniel seemed to have been in overall command of the operation. 1st Lt. James F. Myers directed the Forty-seventh Pennsylvania contingent, and Ensign J. H. Jenks led the naval unit. Daniel's second in command was Lt. Zachariah Brown, a Mississippian who had deserted the Confederate army and enlisted in the Florida Rangers with six comrades at Key West.[20]

The unit left Useppa Island on Christmas Eve, proceeding up the Myakka River for "the purpose of capturing a small number of 'cowboys' who were engaged in driving cattle to the rebel army." (Florida cattlemen have traditionally referred to themselves as "cow men," or "cow hunters," but never as "cowboys.") Daniel also had "instructions to explore the [surrounding] country." Following those directions, the Federal raiders proceeded up the Myakka, keeping the *Rosalie* on their left for added protection. On Christmas night, the bluecoats spotted Confederate signal fires along the Peace River, but the Rebels did not engage the enemy until two nights later. Little, if any, damage resulted from the area's initial "engagement."[21]

Five nights later, on 30 December, at about 4:30 a.m., Rebel cowmen, mounted and on foot, assaulted the camp of the Florida Rangers. According to Union reports,

> Between four o'clock and 4:45 on the morning of December 30, Jenks's pickets discovered figures moving through the tall grass toward camp in the darkness. The pickets challenged them and received the reply they were Daniel's men [who had become

---

[19] Buker, *Blockaders, Refugees, & Contrabands*, 118; Dillon Jr., "Little Affair," 318.

[20] Buker, *Blockaders, Refugees, & Contrabands*, 119–21; "Interesting from Key West," *New York Herald*, 28 January 1864, 1; *ORN*, ser. 1, vol. 17, 610–13; *ORA*, ser. 1, vol. 35, pt. 1, 460–61.

[21] Ibid.; Dillon Jr., "Little Affair," 319–22.

> separated from Meyers's group].... At this time the approaching party, estimated at thirty to forty men, rose in a semi-circle and opened fire with shotguns and Colt revolving rifles, fifteen yards from our picket line. Fire was especially heavy on the Federal right, as the attacking Confederates attempted to cut the Union sailors...off from their boats.

The *Rosalie* opened fire with her artillery, raining "shrapnel and canister fire" on the Rebels until Meyers's squad could take shelter on the gunboat. Only one Union sailor was wounded—a minor leg wound. Union estimates reported Rebel casualties as six to eight killed and many more wounded. Meanwhile, Daniel's contingent, which had indeed gotten separated from their comrades, wandered along the Myakka until they located a small blockade-runner loaded with four bales of cotton, which they commandeered. The last of the Yankee raiders finally arrived at Useppa Island on 1 January 1864.[22]

This admittedly minor affair might have been quickly forgotten, but the Federals discovered that Zachariah Brown, Daniel's second-in-command, had led the Rebel assault on the Union camp. According to the *New York Herald*,

> During the earlier part of the night six of the refugees headed by Brown...deserted. It was now apparent that Brown was brought in three months ago by the Blockading vessel, and voluntarily took the oath of the United States government, [though] he was really a rebel spy...[whose] sole purpose [was] of obtaining all the information he could for the rebels, and to embrace the opportunity to desert.

Another Northern newspaper suggested, "[t]he treachery in this case was deliberate, and will make our [United States] troops cautious of how they trust rebel refugees."[23]

Shortly thereafter, Gen. Woodbury named Henry A. Crane to replace Daniels as the leader of the Second Florida (US) Cavalry. Both Rodney Dillon and Dr. Buker both indicate that the Brown incident had nothing to do with the change of leadership. Buker noted, "Even before he [Woodbury] knew the results of Captain Daniel's expedition, he [had already] decided that

---

[22] Ibid.

[23] "Interesting from Key West"; "Miscellaneous," *Springfield [MA] Republican*, 29 January 1864, 2.

Crane should lead his refugees." Dillon went a step further. Gen. Woodbury believed "that the original refugee commander, Enoch Daniels, though an able scout, was not the right man to lead a large contingent of troops." He further reported that Crane was "well-known and popular among the people of Lower Florida."[24]

HERNANDO COUNTY—BAYPORT AND CRYSTAL RIVER

The US Blockading Squadron had been keeping a close watch on Bayport for several months, and in September, the USS *Two Sisters* spotted a large blockade-runner and several smaller vessels on the river near Bayport. Acting Master C. H. Rockwell, of *Two Sisters*, attempted to send a landing party ashore, but the tides and contrary winds prevented them from reaching the village. The Southerners, however, apparently decided to end their operations at Bayport, burning both the ships and adjacent warehouses. A South Carolina newspaper admitted the setback: "The Confederate troops in the vicinity were small in number and deeming their fire insufficient to save the property at the port, set fire to several vessels lying at the wharf and fell back. Mr. [Thomas Henry] Parsons finding it impossible to save the goods at the warehouse, set it on fire, and it was speedily consumed." Rockwell celebrated the Rebel defeat, bragging, "We arrived off Bayport the next day, and had the pleasure of witnessing the destruction of the steamer, and a large warehouse, apparently containing cotton, and of participating in a bloodless victory, won from the fears of the enemy."[25]

In late 1863, a new menace surfaced along the Gulf Coast. The tightening of the Federal naval blockade, combined with increasing depredations by marauder bands, throughout the sparsely settled region from Cedar Keys to Tampa Bay, brought severe hardship to area residents where "bands of looters roamed freely, inflicting damage wherever they went." One source compared the marauders to "pirates—they robbed all ages and sexes." The son of a member of the Cow Cavalry recalled that his father "hated the deserter [bands] more than the Yankees because of all the damage they caused." Gen.

---

[24] Dillon Jr., "Little Affair," 327–28; Buker, *Blockaders, Refugees, & Contrabands*, 121–22.

[25] *ORN*, ser. 1, vol. 17, 560–61, 579; "Blockade Runner Captured," *Lowell [MA] Daily Citizen and News*, 10 November 1863, 2; "War Items & Incidents," *Salem [MA] Register*, 16 November 1863, 2; "The Yankees in Florida," *Charleston [SC] Courier*, 25 September 1863.

Finegan noted that reports of the marauders' depredations "had a demoralizing influence on men still in the [military] service."[26]

At about the same time that Bayport was under attack, an incident occurred further up the Gulf Coast, in modern Citrus County, that revealed a little-remembered Union experiment in the war. As the Third Occupation of Jacksonville had amply demonstrated, several Federal officers had begun enlisting Black men into colored regiments. Less well known, however, "at times invading Union armies [also] recruited black irregulars. This was the case on the east coast of Florida in early 1863, as Union troops stationed at Beaufort, South Carolina raided [plantations] and encouraged [other slaves to] escape." They received arms, supplies, and encouragement from the Yankee commanders, but the bluecoats had soon put an end to that practice. The military leaders concluded that trained Black men, under capable commanders, could accomplish far more than scattered, poorly equipped irregulars.[27]

One example of the Yankees using Black partisans occurred in northern Hernando County [modern Citrus County] during late August or early September 1863. They apparently were landed in small boats from a ship out of Cedar Keys and raided inland to the area around present-day Lecanto. The Rebel leaders quickly dispatched Capt. Sam Hope with orders to kill or capture the Negro brush fighters. In his description of the operation, Hope wrote,

> In my report to you on the 1st inst., I stated that I succeeded in capturing the boat [used by the Black partisans] and had the negroes cut off so as there would not be much doubt as to getting the negroes...[who had carried out the] raid on Mr. King's plantation. On the morning of the second day I took the trail from where I fired on them the evening previous. They [the tracks] led off towards the mouth of the Withlacoochee River, edging the coast as near as they could for tide [tidal] creeks, & etc. About 4 ock., in the evening we discovered one [of the guerrillas] in a cedar tree looking out [toward] an island. They discovered us about the same time, we being [exposed] in an open marsh. Here they seemed to have separated only two being together. After chasing them about

[26] Ivey, "Accidental Pioneer," 36; Buker, *Blockaders, Refugees, & Contrabands*, 84–85.

[27] Berlin, et al., *Destruction of Slavery*, 805–806. Spelling and punctuation throughout this long passage have been corrected for readability.

> two miles through the saw grass we came up [with]in gun shot [range] of them. We began to fire at them, and they returned fire very cool and deliberately, but soon we got in close range and killed them. One of these negroes was recognized by some of my men as belonging to Mr. Everett who lives near here, which ran away from him about nine months ago. He was styled Captain of the [Negro] party.... Myself and my men being completely tired down for want of water, as we had to go back to camp until the next morning when we took the same trail and followed [the route as on the] third day. About the same time in the evening...we came up on two more...and succeeded in killing both of them, but I am under the impression that we killed or wounded the other three on the first day. I could not get any information from either of the four that was killed, as they were dead.[28]

Hope learned from "an old Negro man of Mr. King's, who ran away the first day and [then] came back home," that the operation began at Cedar Keys. The purpose of the mission seems "to have been to encourage slaves in the area to leave their masters [and enlist in the Federal army], in addition to reconnaissance." A ship from the East Gulf Blockading Squadron had left from Sea Horse Key and dropped the Black irregulars near Crystal River. The gunboat then departed to check the Suwannee River for blockade-runners, but the Union sailors planned to return to pick up the guerrillas and any new recruits.[29]

St. Augustine

In mid-1863, a reporter for a Massachusetts newspaper asserted, "From Florida it is learned that there are no armed rebels east of [the] St. Johns River." While that amounted to a gross misrepresentation of the facts, it appears that the state of siege that had gripped Northeast Florida had gradually weakened. In September, a Union soldier described life in the Ancient City in glowing terms. "We have," he wrote,

> an abundance of excellent fruits: oranges, lemons, figs, pomegranates, limes, grapes, bananas & c., fresh from the gardens of the old Spanish city. It is a remarkably healthy place, much resorted

---

[28] Ibid.
[29] Ibid.

> [prewar] to by invalids during the winter months. We have three large churches...[and] plenty of white people and an abundance of black, yellow, and all colored folks: a steamer with the mail [arrives] once in two weeks.

Another bluecoat reported they also had "a theatre is in full operation...[and it is] very popular with the townspeople."[30]

L. D. Stickney, Eli Thayer, and their fellow Unionists were now in complete control—at least, inside the city. From 1861 through 1864 the US Congress had enacted a series of "Captured and Abandoned Property Acts." The March 1863 version allowed Federal officials, among other provisions, to seize property owned by individuals serving the Confederacy and sell it to pro-Union buyers. Stickney, the amoral opportunist, in his capacity as tax commissioner for Florida, in December 1863, held a "sale of rebel property for taxes in the old town of St. Augustine." For obvious reasons, "few of the owners, or their representatives, were present at the sale." Buyers from Massachusetts, New York, Illinois, Wisconsin, and other Northern states purchased the land at cut-rate prices. Properties not sold "passed...into the hands of the general [Federal] government." "The plans for leasing plantations," the reporter averred, "is working with similar results, by bringing the landed estates lately possessed by rebels into the hands of loyal occupants and under the culture [cultivation] by free, compensated labor."[31]

Not all was rosy in the Ancient City. A soldier, in a public letter, griped, "The government feeds about fourteen hundred residents of this place, most of whom are too indolent or proud to work." Additionally, a number of wounded soldiers, from the bitter combat around Charleston, began pouring into town. The Union established a convalescent camp at the old town, and more casualties soon flowed into Northeast Florida.[32]

---

[30] "By Telegraph from Port Royal," *Worchester [MA] National Aegis,* 2 May 1863, 3; "Letter from St. Augustine, Florida," *Lowell [MA] Daily Citizen and News,* 15 September 1863, 2; [no title], *New Haven [CT] Palladium,* 7 December 1863, 2.

[31] "Progress of Events," *Lowell [MA] Daily Citizen and News,* 31 December 1863, 2; Schafer, *Thunder on the River,* 175–76; R. Taylor, *Rebel Storehouse,* 134–35.

[32] "Connecticut Regiments in Gilmore's Department," *New Haven [CT] Palladium,* 4 December 1863, 1; "A Letter from St. Augustine, Florida," *Portland [ME] Daily Eastern Argus,* 8 December 1863, 1.

The lull in guerrilla activity around St. Augustine emboldened Union troops to increase their activities beyond the city's walls. A Federal soldier boasted that his unit, in early December, had gone into the countryside to look for meat for the garrison. He wrote, "Seven companies of our regiment under the command of Major Hooper, made a little excursion thirty miles into the interior last week and returned with fifty head of cattle, taken from such [citizens] as did not recognize the authority of the United States, and would not swear by the stars and stripes." The number of beeves soon grew to two hundred, as various Northern newspapers reprinted the account, but the bluecoats cared little regarding this journalistic puffery, as they feasted on steak and roast.[33]

Col. Francis A. Osborn, commanding the Federals at St. Augustine, made it a habit to keep a close eye for signs of guerrilla activity east of the St. Johns River. He sent out scouts constantly and took precautions to make sure his men did not get caught in an ambush. On 30 December 1863, he received intelligence that a large band of Rebel irregulars might be east of the St. Johns. Osborn took the reports seriously. He wrote that on hearing "rumors that some [Confederate] cavalry were expected to cross the river very soon [looking] for conscripts and deserters, and I accordingly increased the guard [for woodcutters being sent to the nearby woodlands] to 30 men, [and] requiring the 20 choppers to carry arms, making 50 men, which…I deemed an ample force."[34]

Capt. J. J. Dickison, with his Co. H, had already crossed the St. Johns the previous evening. He had a total of about seventy men with him, including a small detachment from Capt. Chambers's Co. C, under the direction of Lt. Samuel Clarke Reddick. Lt. William McCardell, Dickison's second in command, led the men from Co. H.[35]

The Unionists had gotten in the habit of going to the same patch of forest (only two miles from the city gates) at about the same time of day, which made an attack easy to plan. Lt. Walker, with men from the Twenty-fourth Massachusetts, accompanied the woodcutters. Walker began establishing a perimeter near an oak hammock, when a small party of graycoats

---

[33] "Letter from the 24th Mass. Regiment," *Boston Herald*, 5 December 1863, 3; Roe, *Twenty-fourth Regiment of Massachusetts Volunteers*, 244.

[34] *ORA*, ser. 1, vol. 28, pt. 1, 722–25.

[35] Ibid.; "An Honorable Disclaimer," *Charleston [SC] Courier*, 14 January 1864, 1.

opened fire from the surrounding palmettoes. Col. Osborn noted the enemy had been "lying concealed in low palmetto shrubs with which the whole country is covered, and which furnishes such perfect concealment that a man might pass within 20 feet of such a party and never suspect its presence." Lt. Walker, a battle-tested veteran, started forming his men in a defensive alignment, when he fell mortally wounded. (Some sources suggest that Walker might have been a victim of friendly fire, when one of his rattled soldiers prematurely discharged his weapon.) At that moment, the rest of Dickison's troop spurred their horses into the Union defenses, attacking the enemy from the front and rear. The woodcutters, forgetting their weapons, quickly turned their animals around, whipping the mules toward the Ancient City.[36]

An article in the *New Orleans Times-Picayune* provided a good summary of the fight near St. Augustine. It reported,

> On reaching a point about five miles from town a sudden fire was opened on both sides of the road, which wounded one or two. Lieut. Walker, of the 24th Massachusetts, who was in command of the guard, attempted to make a successful fight, but to no avail. The rebels were a hundred [actually, seventy] strong, and had the advantage of surprise. The result of the affair was that Lieut. Walker was mortally wounded, and a private named Burns…[was also] mortally wounded, and twenty-five taken prisoner…. On information being received of the affair, a strong force was sent to overtake the enemy, but the latter being mounted escaped…. The rebels treated our wounded…with great kindness. They made no attempt to rob them, but left watches and purses on the bodies of the dead. Dickerson [Dickison] placed his own saddlebags under Lt. Walker's head and expressed his regret that he was unable to make him more comfortable.[37]

One final event occurred before the raid near the Ancient City faded into obscurity. Gen. Finegan sent Capt. James F. Tucker, commanding Co. D, of the 6th Florida Battalion, to escort the Twenty-fourth Massachusetts

---

[36] *ORA*, ser. 1, vol. 28, pt. 1, 722–25; "An Honorable Disclaimer"; "Lieut. Walker of the Mass 24th Wounded and a Prisoner," *Boston Evening Transcript*, 19 January 1864, 1.

[37] "Raid Near St. Augustine, Florida," *New Orleans Times-Picayune*, 30 January 1864, 2.

prisoners to Charleston. As they passed through Savannah, the local newspaper gave Tucker and his unit the acclaim for capturing the bluecoats. Tucker, however, insisted the newspaper print a retraction, giving Dickison's and Chambers's men credit for taking the prisoners.[38]

[38] "An Honorable Disclaimer."

Chapter 6

"Numerous Infernal and Barbaric Acts"

(January–July 1864)

OLUSTEE CAMPAIGN

The last six months of 1863 proved to be the calm before the storm. During the early weeks of 1864, the Union Army began making every effort to curtail shipments of Florida beef to the Rebel armies in Georgia, South Carolina, and Virginia. With strong support from Maj. Gen. N. T. Banks, commanding the Department of the Gulf, a Federal cavalry force composed of Florida Unionists and deserters from the Rebel armies initially proved quite effective. Banks, a cautious officer, advised his subordinates in Florida that goal of halting the flow of South Florida cattle "is an advantageous and proper one, but great care should be taken to avoid any surprise by the enemy."[1]

Adding to the Rebels' concerns, all across the state large numbers of Confederate defectors and conscript evaders formed gangs to conduct marauding raids on plantations, small villages, and isolated farmsteads. The danger became so severe, in February 1864, that it even threatened Florida's chief executive . Gov. John Milton received a warning telegraph that deserter bands in West Florida planned to kidnap him as he returned to Jackson County for a vacation. According to the tip, the marauders' plot involved capturing him and turning him over to one of the Union blockading ships in the Gulf. As a result, Milton "decided to remain in Tallahassee to avoid becoming a prisoner or a casualty of the deserters."[2]

The Richmond government, rather than easing up on conscription, began cracking down on the males who refused to answer the call to join the military. In a 26 January letter to James A. Seddon, Confederate secretary of war, Gov. Milton predicted, "[T]he wave of indignation concerning impressment will drive even greater numbers [of Florida draft evaders] to avoid military service if the evils of the system are not immediately corrected." He

---

[1] Proctor, *FL100*, 4 January 1864.

[2] Ibid., 3 and 4 February 1864.

explained that most of the runaways left the army to "save their families from starvation." When forced to choose between their wives and children, or the Confederate States of America, many men picked the former.[3]

The Federals' initial Florida operation—the Olustee Campaign—had as much to do with politics as it did with military objectives. With the near deification of Pres. Abraham Lincoln by many modern Americans, it might seem shocking that "Honest Abe" seriously doubted if he could win the 1864 presidential election. His Democratic opponent was war hero Gen. George B. McClellan, and there was anger in the North, where many voters despised the Emancipation Proclamation and the idea of granting African Americans the same rights as Whites. There was also a splinter group within the Republican Party, headed by Salmon P. Chase, secretary of the treasury, who thought that Lincoln had not done enough for the South's Negro slaves.

Lyman D. Stickney and several other Machiavellian opportunists, encountered this volatile situation when they arrived in Washington, D.C., preaching the old "gospel" that the most White Floridians longed to return to the Union—"their first love." Stickney and his cohorts suggested that if the Federals could win a major victory on Florida soil, the "President's Tenth" could be invoked, giving Lincoln, or Chase, a substantial political advantage. A recent historian explained,

> The proclamation to which Stickney referred was Lincoln's Amnesty Proclamation of December 8, 1863, which set forth the ten percent plan of reconstruction. According to this plan, citizens [in seceded states] were to take the oath of allegiance to the United States in order to become voters. When enough had done this to comprise ten percent of the state's voters in 1860, Lincoln would recognize a state government of their creation.

Under this bizarre policy, a state, actively fighting the bluecoats as a part of the Rebel Confederacy would still be allowed to cast votes on Union legislation. This absurd scheme would put Florida—with armies in the field actively trying to defeat Lincoln's government—on equal footing with New York or Massachusetts.[4]

---

[3] Ibid., 26 January 1864.

[4] Futch, "Salmon P. Chase," 188. See also, McMurray, "President's Tenth and the Battle of Olustee," 13–25. The best account of the Byzantine maneuvering by Stickney and his cohorts can be found in Coles, "Far from Fields of Glory."

Meanwhile, the Federals' planned invasion of Florida slowly moved forward. Many Northern military leaders had little faith in the project, grumbling that "politics permeated every aspect of the Florida campaign." Gen. Henry W. Halleck, general-in-chief of the Union armies, considered the invasion of Florida nothing less than a disaster waiting to happen. If the Federals successfully defeated the Rebels, he averred, the state would merely swallow up "our troops in garrisons to occupy the place, but have little or no influence upon the progress of the war." (Two months later, Pres. Lincoln replaced Halleck with Gen. Ulysses S. Grant.) Adm. John Dahlgren, commanding the South Atlantic Blockading Squadron, also opposed the invasion, but followed orders, providing transport from South Carolina to Jacksonville. At first, Gen. Quincy Gilmore expressed support for the operation, but he, too, quickly soured on the proposed campaign. Seemingly, only Lincoln and Brig. Gen. Truman Seymour, who would lead the Union army, remained true believers.[5]

Seymour had proven himself a brave man on many of the Civil War's bloodiest battlefields. He survived the 1861 bombardment of Fort Sumter and had fought in several gory clashes, including Antietam, South Mountain, and Second Manassas. He had most recently led a "bloody frontal assault on entrenched Confederate troops at Battery Wagner, Morris Island, and during the first attempt to capture Charleston [South Carolina] in July 1863." Severely wounded, he had only returned to the army in December 1863. Among the common soldiers, Seymour possessed a reputation as a "sometimes rash commander, who often succeeded in battle, but at a heavy cost."[6]

On the morning of 6 February, the armada, consisting of more than forty vessels, departed Hilton Head. The ships carried a force of seven thousand men, including White and USCT infantry, as well as cavalry and artillery. They arrived the following day at Jacksonville, which appeared dismal. "Jacksonville is 'pathetically dilapidated,' a mere skeleton of its former self, a victim of the war," a journalist reported. "Struggling winter weeds grow in the streets, and the remains of burned houses give a grotesque, Godforsaken, and dreary aspect to the town."[7]

---

[5] *ORA*, ser. 1, vol. 35, pt. 1, 279; Schafer, *Thunder on the River*, 176–78; Coles, "Far from Fields of Glory," 31–42.

[6] Schafer, *Thunder on the River*, 178; "Brigadier General Truman Seymour, U. S. A."

[7] Schafer, *Thunder on the River*, 178–79; Proctor, *FL100*, 7 February 1864.

The Yankees did not even get a chance to land before they drew fire from a small guerrilla band. A member of the Fortieth Massachusetts (Mounted) Infantry related, "As soon as the steamers *Gen. Hunter*, *Maple Leaf*, and *Charles Houghton*...attempted to moor at the wharf, they were met by a volley from some mounted rebels. The mate and pilot of the *Maple Leaf* were both seriously wounded, and others slightly." A Bay State cavalry unit, joined by the veteran Fifty-fourth USCT Infantry, quickly disembarked from the vessels and began their pursuit of the fleeing graycoats.

> The detachment immediately started, traveling partly on [along] the railroad and then through some woods, in pursuit of the mounted rebels. We succeeded in capturing five of them. When [we had gone] about seven miles, Major [Atherton] Stevens came suddenly upon a signal station, in full operation. The rebel party, not knowing the Yanks were so near, we succeeded in capturing five at the station, with all implements. After destroying their buildings, the Major went still further, captured four more [Confederates], and brought them all to headquarters.

All of the captives refused an offer to take the oath of allegiance.[8]

Early the next morning (8 February), a Federal expedition left the River City. Col. Guy V. Henry planned to strike as far inland as his unit could safely proceed. An 1861 graduate of West Point, one of Henry's superior officers reported, "Probably there is no officer in the [US] Army more indefatigable in the performance of his duty than...Guy V. Henry." Before he began his raid, Henry wisely had his men exchange their single-shot muskets for a Connecticut regiment's Spencer repeating carbines. Henry's reconnoitering party consisted of elements of the Fourth Massachusetts Cavalry, Fortieth Massachusetts (Mounted) Infantry, and Elder's Artillery. To insure that his units would not get bogged down with prisoners and the spoils of war, two infantry regiments, including the Fifty-Fourth Massachusetts (USCT), followed in the raiders' wake to take charge of any "baggage" that might otherwise slow down Henry's horsemen. In another example of Henry's military acumen, each of his units traveled separately with a "loyalist" guide at the head of each column, who "knew every road and bypass in the country." In

---

[8] Schafer, *Thunder on the River*, 179; "Excellent Fighting by Massachusetts Troops," *Boston Herald*, 22 February 1864, 1.

that way, "the Federals were able to make such swift dashes through the night."[9]

Camp Finegan, eight miles west of Jacksonville, would be Col. Henry's first target. The Federals hit it after dark, and a Confederate soldier wrote, "The movement of the enemy was so sudden that they came within four hundred yards of our camp before they were discovered." The astonished graycoats scrambled to form a defensive line, but it proved too little, too late. So complete had been the surprise that a second Rebel soldier wrote, "Our troops were soon drawn up in line, but before they could move off were flanked, [to the] right and left, and about one hundred and fifty of them captured...[N]ot one gun was fired by us." Capt. Joseph L. Dunham, commanding a Florida battery, was on sick leave, but his men also lost their guns, caissons, and most of their horses. Luckily for them, in the inky darkness a number of the Confederates managed to escape. Henry's fast-moving Yankees halted just long enough to set a guard over the prisoners and the captured plunder, then spurred forward to their next objective.[10]

Henry's raiders hastened on toward Baldwin, a small but important village just a few miles west of Camp Finegan. The state's two major railroads—the Atlantic & Gulf Central Railroad (running from Jacksonville to Tallahassee to Apalachicola) and Florida Railroad (running from Fernandina to Cedar Keys)—intersected at Baldwin. Scattered around the terminal were warehouses containing all manner of valuable military supplies. A local historian noted, "The Confederacy...[had] stored in the warehouses...[c]amp equipment, cannon, accouterments, forage, cotton thread, cotton sheeting, rice, molasses, blankets, salt, hides, sugar, flour, turpentine, and such." In addition to those goods, the Yankees also captured a boxcar containing a three-inch rifled artillery piece and its caissons. Despite the hamlet's importance to Florida's Confederates, Henry's bluecoats occupied it without firing a shot.

---

[9] Erlandson, "Guy V. Henry," 20–22; Burdett, "Military Career of Brigadier General Joseph Finegan, CSA, of Florida," 19. Henry afterward served in the wars against the Plains Indians, during the Spanish-American War, and ended his service as governor of Puerto Rico.

[10] "Latest from Florida," *Macon [GA] Telegraph*, 16 February 1864, 2; "The Raid into Florida," *Charleston [SC] Courier*, 16 February 1864, 1; "Repulse of the Enemy in Florida," *Charleston [SC] Mercury*, 15 February 1864, 1; Schafer, *Thunder on the River*, 179–84.

Again, the Unionists "remained a short time" and then galloped toward Barber's Ford.[11]

Barber's Ford, on the South Prong of the St. Marys River (near modern MacClenny), offered the Rebels an excellent place to delay Col. Henry's column. Gen. Finegan had earlier directed Maj. Robert Harrison to bring two companies of the 2nd Florida Cavalry to join the Confederate force being assembled near Lake City. Seeing a golden opportunity to "check their [Henry's raiders] progress for several hours, Harrison deployed his men along the west bank of the stream."[12]

Thus far, Henry's troops had met almost no resistance, and they expected none at the bridge over the St. Marys. A Union soldier recounted,

> The bugles sounded the call to mount, and...the Massachusetts battalion...started forward and entered a small defile leading through thick impenetrable underbrush and pine trees to the bridge across the St. Mary's. The platoon of four men of the cavalry which had the advance had just passed a sharp turn in the road, and had approached close to the bridge, without anticipating an attack, when a half a dozen reports were heard, and three or four fell from their saddles, shot by a rebel force ambushcaded in a strong position beyond the stream. The column immediately came to a halt when the presence of the enemy was made known by the explosion of their guns, and the advance company pushed forward under Captain [Richmond] Webster, to feel and ascertain their position. It was received with a sharp volley of musketry, which dropped out of their saddles several brave men. The fire was instantly returned, but with little effect as the enemy were concealed behind bushes and stumps, from which they could use their guns with deadly effect.[13]

The Bay State soldiers charged Harrison's position twice, and each time they fell back with more empty saddles. The Union soldiers who made it to the river, soon realized that they could go no further. "[I]t was discovered," a Massachusetts trooper wrote, "that the river at that point was not fordably

---

[11] Adams, "The Baldwin Story: Baldwin, Florida"; Schafer, *Thunder on the River*, 180; *ORA*, ser. 1, vol. 35, pt. 1, 331.

[12] Schafer, *Thunder on the River*, 181; *ORA*, ser. 1, vol. 35, pt. 1, 330–33.

[13] Ibid.; "Excellent Fighting by Massachusetts Troops."

[fordable] and the attempt was relinquished." Col. Henry decided to add some weight to his units' assaults: "Elder's battery was placed in position on the crest of a little hill in front of Barber's house, which gently sloped down to the riverbank, and the cavalry and mounted infantry were placed on either side in line of battle to support it." Meanwhile, a dismounted company spread out to the north and south "where the river makes a sudden turn." From there, they opened an enfilading fire on the men of the 2nd Florida Cavalry, and the artillery fire convinced Maj. Harrison it was time to fall back toward Lake City. Casualties for each side likely numbered twenty men killed and wounded.[14]

Henry's battered strike force pushed west toward Sanderson, where they discovered "the central portion [of the village] in flames." Both the men and horses were exhausted, and Col. Henry decided to give them a brief respite. They waited until the infantry caught up with the "flying column," and then his mounted men continued through the night toward Lake City.[15]

Gen. Finegan had established a line of breastworks a few miles east of Lake City, where the majority of his force waited for the Federal raiders. As soon as the Union flotilla had arrived at Jacksonville, the Irish general began sending urgent orders for his widely scattered companies to hasten to the Columbia County town. He also began badgering Gen. P. G. T. Beauregard, commanding the District of South Carolina, Georgia, and Northeast Florida, to send him some veteran troops. By the morning of 11 February, he had managed to assemble a force of about 490 infantry, 100 cavalrymen, and two pieces of artillery. Col. Henry had lost some men at Barbers, and after several days on the road there would have had a few stragglers, but his force likely equaled, or even exceeded, the number of Finegan's graycoats.[16]

Early on the morning of 11 February, Henry's Union skirmishers advanced through a heavy fog, feeling for the Rebel position. Finegan had a skirmish line, comprising men from the 1st Georgia Regulars, in advance of the main Rebel defenses. Lt. John Porter Fort commanded the Southern skirmishers. A Georgian on the main line wrote, "[W]e could see our boys and the Yankees as they fought from tree to tree, flanking first on one side and then the other." Fort confirmed the account, adding, "It was a foggy morning and the enemy approached within seventy-five or a hundred [yards] before

---

[14] Ibid.

[15] Ibid.

[16] Coles, "Far from Fields of Glory," 53–55.

we perceived each other...I was instructed to try to draw them near our line. Both sides commenced firing. Soon the mist rose. The enemy, seeing our line of battle, retreated with haste." In his report to Beauregard, the Irish general accurately deduced that the enemy's main purpose involved verifying Confederate strength in North Florida. "At 9:30 [a.m.]," he reported, "the enemy advanced upon us with a force estimated to be 1,400 mounted infantry, and five pieces of artillery. Here they opened on us, fighting as infantry, and skirmished heavily with my advance line. Discovering my position and its strength, and probably presuming my force was larger than it was, they retreated to Sanderson, and thence to Barber's."[17]

Though Henry retreated from Lake City, he had accomplished his purpose. He had discovered the strength and position of the Confederate forces in North Florida, as well as capturing several cannons, at least 150 enemy soldiers, and perhaps a half a million dollars in vital war supplies. A Union report summed up the raid, stating a bit too optimistically, "The success of the expedition is beyond question...[and] this raid will cut off the chief cattle supplies of the rebels." (Forgotten in the rejoicing, Henry had failed to accomplish one of his secondary objectives—the destruction of the railroad bridge over the Suwannee River, at Columbus, Florida.[18])

While Col. Henry retreated toward Jacksonville, Capt. George E. Marshall, of Co. E, Fortieth Massachusetts Mounted Infantry, took a force of about sixty men south to Gainesville. The county seat of Alachua County sat astride the Florida Railroad, and the Confederate government had constructed several warehouses there to house cotton and military supplies. Marshall had "skirmished all night and reached the place on Sunday morning, February 14, at 2 A. M." Capt. Marshall displayed his considerable military skill, setting out "sentries and pickets."[19]

The next morning Marshall and his troops explored Gainesville, finding 167 bales of cotton and about thirty-five Black men who indicated their desire to follow the bluecoats to freedom. One of the Negroes informed Marshall that a Rebel soldier had managed to slip through the Yankee lines and would likely be back soon with the Confederate cavalry. The Bay State

---

[17] Fort, *Memorial and Personal Reminiscences*, 23; McMurray, ed., *Footprints of a Regiment*, 123–24; Coles, "Far from Fields of Glory," 53–57.

[18] Ibid.[**Coles?**]

[19] "The Late Attack at Gainesville, Florida," *Boston Evening Transcript*, 1 March 1864, 4; "General Gilmore's Departed," *Philadelphia Inquirer*, 17 February 1864, 2.

captain immediately set his men to work constructing barricades of cotton bales. Armed with repeating rifles and protected by those breastworks, the bluecoats felt confident that they could hold their own against any force of Rebel irregulars.[20]

Lt. E. A. Davis, with Co. C, 2nd Florida Cavalry and a few men from Dickison's Co. H., arrived during the morning, determined to drive the Yankees from Gainesville. (Capt. W. E. Chambers, who had previously displayed considerable skill as a bush fighter, had been forced to resign due to a serious illness.) A Union soldier reported, "The rebel soldiers were soon heard thundering down the road. Captain Marshall enjoined his men to hold their fire until they [the Confederates] should be close to the breastworks." Davis seemingly rode headlong into Gainesville without scouting the situation beforehand. A Union participant in the fight said the foremost Rebel horsemen were close enough to leap the smaller obstructions when a seven-fold volley was poured into them from the bluecoats' repeating rifles. Instantly wheeling, the rebels tried a flank movement, when a terrible enfilading fire ripped through their column, with every man of the national force firing seven shots at the astonished Rebel troopers. A total rout was the result. The frightened horses of the dismounted rebels came vaulting over the stacked cotton bales. The groans of the wounded left by their comrades mingled with the shouts and cheers of the Union soldiers. The men later boasted that they had whipped Dickison's men in a standup fight and decimated his unit.[21]

Marshall's bluecoats, ensconced behind breastworks and armed with repeating rifles, had indeed scored a solid victory over the Florida guerrillas. They apparently erred, however, in whom they beat and the number of casualties. (During the time of the First Battle of Gainesville, Capt. J. J. Dickison and most of his men were on a special assignment, with Lt. Col. Brevard, in South Florida. Having been sent to the area east of Tampa to protect the flow of cattle to the Confederate armies in Georgia and South Carolina, a disgusted Dickison later wrote: "After a march day and night of 575 miles with little rest, they [he, Brevard, and most of his company] were too late by twelve hours to take part in the battle [of Olustee].") Most modern historians have accepted the Federal claim that they inflicted almost forty casualties on Davis's unit, but existing records list one dead and three wounded in the February attack on Gainesville. A comrade described the dead man, John R.

[20] Ibid.; Stephens, *Withlacoochee Notes*, 28–29.

[21] "The Late Attack at Gainesville, Florida."

Worthington, of Co. C, "as brave and gallant a soldier as ever wore the gray." Seriously wounded were Samuel C. Reddick and Benjamin P. Rouse, of Co. C; William Hall, of Dickison's unit, apparently received a lesser wound. Including minor wounds, the total casualties likely numbered about ten men. (To somewhat ameliorate the losses to the two units, Gainesville teenager W. B. Turner enlisted in Dickison's unit while they camped in Alachua County.[22])

Gen. Beauregard had correctly deduced that Henry's raid, and the presence of a large Union force at Jacksonville, served as precursors of a major expedition into the North Florida interior. Finegan had less than twelve hundred soldiers at his disposal, so Beauregard began dispatching reinforcements to Florida. He ordered Brig. Gen. Alfred Holt Colquitt's veteran unit—the 6th, 19th, 23rd, 27th, and 28th Georgia Regiments—to Florida. Colquitt's men had fought in some of the Army of Northern Virginia's hardest battles, until they were exiled to the Deep South, when a minor infraction at Chancellorsville drew the ire of Gen. T. J. "Stonewall" Jackson. Col. George P. Harrison's units—the 1st Georgia Regulars, 32nd Georgia, 64th Georgia, Bonaud's [Georgia] Battalion, and Col Charles Hopkins's 1st Florida (Special) Battalion—had much less combat experience than Colquitt's units. Col. Duncan Clinch's Fourth Georgia Cavalry and some companies of the 2nd Florida Cavalry would serve as Finegan's mounted arm. All hurried south, and by the day of battle, the Irish general would have about five thousand soldiers.[23]

Despite his lack of battle experience, Finegan chose his position wisely. He anchored his breastworks "on Ocean Pond to the north and impassible swamps to the south." Finegan averred it to be "the only defensible position in that part of the state." The battle would be fought over typical North Florida terrain—pine trees scattered across flat, open, stretches of wiregrass, and dotted throughout with the ubiquitous dog fennels and low palmetto bushes. The pines that covered the area were not dense (growing close together) but far enough apart that the combatants could see for good distances, and passage through this forest was not difficult for artillery or cavalry.[24]

---

[22] Ibid.; Evans, ed. (and J. J. Dickison), *Confederate Military History* (Florida), 11:55; Hartman and Coles, *BRF*, 4:1407–557.

[23] Coles, "Far from Fields of Glory," 99–106; Hillhouse, *Heavy Artillery & Light Infantry*, 65–66.

[24] Hillhouse, *Heavy Artillery & Light Infantry*, 65–66.

Early on the morning of 19 February, Seymour's Federal army began marching toward Olustee. The Union commander had approximately fifty-five hundred infantry, cavalry, as well as sixteen artillery pieces. (Approximately two thousand bluecoats remained at Jacksonville to guard the town.) Col. Henry's Mounted Infantry led the advance though some sources claimed they "raced on [around the battlefield] to destroy the railroad bridge over the Suwannee River."(Since the span remained in use, we can assume the writer was mistaken.)[25]

Florida's only major battle, the 20 February fight at Olustee, had little to do with guerrilla warfare. Colquitt, the Confederate field commander, had fought with the armies of Lee and Beauregard and he used the stratagems he had learned in the fighting in Virginia and South Carolina. Virtually none of the Florida troops had ever been in a major engagement, but they showed plenty of courage and determination at Ocean Pond. Capt. Winston Stephens boasted,

> Our men [went into] the fight against their [the bluecoats] immense numbers and we drove them as prettily as you ever saw anything done.... The old Georgia troops say they have never seen better fighting done any where but say our [Florida] boys did better than any men they ever saw. They say they thought the Fla boys would not fight here but now they say they will go further than they will go, the fact is men cannot fight better than ours did.

While Stephens might be forgiven his state pride, the Florida soldiers—especially the 6th Florida Regiment and 1st Florida Special Battalion—had contributed a great deal to the Rebel victory. The two Florida infantry units lost a total of 132 men, killed, wounded, and missing.[26]

Several officers who had previously gained a measure of fame as guerrilla warriors, including Capt. Sam Hope and J. W. Pearson, fought at Olustee. (Hope had nearly missed the engagement. Following the sharp fight with Henry's Federals at the St. Marys River, Hope and his men found themselves

---

[25] Schafer, *Thunder on the River*, 185–87; Coles, "Far from Fields of Glory," 116–17.

[26] Blakely, et al., *Rose Cottage Chronicles*, 320; Baltzell, "Battle of Olustee," 199–223. Several other studies are available for this engagement, including Coles, "Far from Fields of Glory"; Nulty, *Confederate Florida*; Broadwater, *Battle of Olustee, 1864*; and Schmidt, *Battle of Olustee.*

surrounded. Rather than panicking or trying to fight their way out of the trap, the Rebels simply took cover in a dense woodlot, while the Union horsemen passed by them along a pair of roads on either side of Hope's hideout.) Hope carried thirty men into Ocean Pond fight and suffered five killed and ten wounded. Pearson commanded the 6th Battalion at Olustee. According to Dr. Knetsch, "The 6th Battalion was stationed on the extreme right of the battlefield near the railroad tracks. 'There it opened a deadly enfilade on the 8th [United States] Colored Troops…inflicting such severe damage as to compel them to fall back in mass confusion.'" (To do justice to those brave Black soldiers, they entered the battle "never [having] been initiated into the mysteries of handling a musket;" and they suffered almost 50 percent casualties before retreating.)[27]

Caraway Smith's after action account averred Clinch's Georgia horse soldiers took a position on the Rebel left, while McCormick's 2nd Florida Cavalry took a similar place on the right. Finegan related, "On two occasions I discovered that the enemy was attempting to cross the railroad on the right of his [McCormick's] regiment and [ordered the Florida Cavalry to] drive them back, which he did effectually." Capt. Winston Stephens concurred. He wrote to his wife after the fight,

> When we got to the infantry line[,] I was thrown to the right flank[,] some times I was in Command…. The enemy pressed us quite hard but our artillery and infantry opened and the boys [horsemen] went to work as men can only work who are in earnest, then the scene was grand and exciting. I felt like I could wade through my weight in wild cats. The 2d Cav was [then] dismounted to fight on foot and I think we did good work[;] we went in with a wild yell and the Yanks and negroes gave way, then we would remount and follow up and we continued that [way] until the fight ended.

Col. Clinch received a wound shortly after the combat began, likely diminishing the effectiveness of the Georgia horsemen. Maj. G. W. Scott, of the Fifth Florida Cavalry, arrived during the middle of the Olustee engagement. Scott's men, however, contributed little to the victory as his men and

---

[27] Hope, "Report of S. E. Hope's Part in a Civil War Incident;" Knetsch, "Forging the Florida Frontier," 35; Dobak, *Freedom by the Sword*, 65–70.

animals were in a "jaded condition...from hard service for the twenty-four hours preceding."[28]

Following the Federals' retreat, Col. McCormick's 2nd Florida Cavalry and Clinch's 4th Georgia Cavalry received orders to pursue the retreating Yankees. Stephens continued his chronicle of events in the engagement's aftermath, relating, "Then [after the Union retreat] we were thrown to the front and we got during the night some 200 yanks that were wounded and not able to keep up with the main body. We were returned this morning after having been in the saddle one day and night without any rest or anything for our horses to eat." In summing up his report on the Olustee fight, Caraway Smith boasted that his cavalry troopers "behaved with the coolness and deliberation of veterans."[29]

Despite the hard-won victory, Beauregard, Gov. Milton, and Gen. W. M. Gardner, commanding a tiny Confederate subdistrict in the Panhandle, unleashed a firestorm of criticism upon Gen. Finegan and Col. Smith for failing to actively pursue the routed enemy. Smith pointed out that his men had floundered through the darkness with horses and men worn to a frazzle by days of skirmishing and fighting. From a perspective of a century and a half, the Florida horsemen seem to have done all in their power to chase down the fleeing bluecoats. Caraway Smith immediately demanded a court of inquiry to clear his name, which apparently was never convened. Newspaper reports also claimed Robert H. Anderson, of the 5th Georgia cavalry, would replace Smith, but that change never happened. Col. Smith survived the charges and fought at the Battle of Natural Bridge in mid-1865, the second largest engagement in Florida.[30]

Despite the backbiting and finger pointing, which seemed endemic in the Southern military and government, the victory at Olustee bolstered the morale of Florida's pro-Confederate citizenry. The *Floridian* issued the following fervent appeal: "We tell the people of Florida to be of good cheer. Don't give way to despondency.... Rally to the defense of your country. Prepare to act a sublime and heroic part in the war...[I]f a Yankee army ever

---

[28] *ORA*, ser. 1, vol. 35, pt. 1, 352–53; Blakely, et al., *Rose Cottage Chronicles*, 319.

[29] Ibid.

[30] *ORA*, ser. 1, vol. 35, pt. 1, 352–56; Coles, "Far from Fields of Glory," 145–47; W. C. Davis, "William Montgomery Gardner," in *Confederate General*, 2:162–63; "The War in Florida," *Richmond [VA] Examiner*, 27 February 1864, 2.

[again] penetrated into the forest and swamps of Florida, it would be a shame if they were ever allowed to escape."[31]

NASSAU COUNTY (JANUARY–MARCH 1864)

With most of the Confederate troops in Florida concentrated near Lake City, the Federals at Fernandina took the opportunity to raid into the Nassau County interior. On 15 February, a three-hundred-man contingent of the Ninety-seventh Pennsylvania Infantry, commanded by Maj. Galusha Pennypacker, landed near Clark, on the Amelia River. While Pennypacker's men marched overland, a flotilla of ships, consisting of the *John Adams*, *Harriet A. Weed*, *Island City*, *Beaufort*, *Nelly Baker*, and a mortar schooner, *Para*, proceeded up the St. Marys to Woodstock Mills and King's Ferry. Maj. T. B. Brooks, of Gen. Gilmore's staff, directed the naval contingent. Upon reaching the two sawmills, they seized "over two million feet of superior yellow pine, well-seasoned, and said to be worth, in New York, one hundred dollars per thousand." A second Northern journal guessed that the haul amounted to an astounding 1.5 million feet of pine lumber.[32]

The raiding party met little in the way of opposition from the Rebels. Company K, of the 2nd Florida Cavalry, had ably protected the Nassau County interior, but they had been ordered to Lake City, and Capt. Frank Clarke had retired in mid 1863, enlisting in a cavalry company in his native Virginia. (Lt. Jesse N. Jones now commanded Co. K.) An unnamed Rebel militia unit left to guard Woodstock and King's Ferry Mills, had apparently been surprised by the bluecoats, and fled without burning the lumber. Another source reported, a "slight skirmish with a rebel picket near the mill at Judge Alberti's was all the opposition we met." Pennypacker claimed that he and his men appeared so unexpectedly that the Confederate "pickets were unable to fire the lumber, as they were ordered to do in case of our approach." Meanwhile, the *Para* showered the woods near Woodstock Mills with mortar

[31] "The War in Florida," *Richmond [VA] Examiner*, 23 February 1864, 2, quoting the *Tallahassee Floridian*.

[32] *ORA*, ser. 1, vol. 35, pt. 1, 359–60; *ORN*, ser. 1, vol. 15, 284–85; "The Florida Campaign," *New York Herald*, 1 March 1864, 1; "Successful Raid Up the St. Mary's," *Port Royal [SC] New South*, 5 March 1864, 1.

shells, increasing the pace of the home guard's flight. The Yankees reported two Union soldiers "very slightly wounded."[33]

In one of those odd events, that seem to happen with considerable frequency in Civil War Florida, a Unionist newspaper reported, "[A] widow named Mrs. Alberty [Alberti] claims ownership of the lumber taken at the Ferry. She resides in the immediate neighborhood, and professes to be a strong adherent to the Union.... Her case will, of course, be investigated." (According to Federal law, the government had to pay full value for property taken from Southern Unionists.) Edwin R. Alberti, who died in 1861, had been born in Philadelphia, and his second wife, Ernestine, hailed from New York. While living in Pennsylvania, they had alienated their neighbors by insisting that the Fugitive Slave Act be strictly enforced. After moving to Northeast Florida, they antagonized their neighbors by sending a mulatto woman they owned to the North to marry a White man. They had also hired gospel preachers from the North to minister to the spiritual needs of their "chattel," but objected to the preachers filling the Black people's heads with abolitionist "nonsense." Despite the family's complicated—almost bipolar—belief system, if Mrs. Alberti could prove her ownership of the lumber—and her fidelity to the Union—she would soon become a very wealthy woman.[34]

JACKSONVILLE

Following the fight at Ocean Pond, Jacksonville settled into a tense equilibrium. The last of the defeated Federals staggered into the River City on the evening of 22 February, and the bluecoats quickly began constructing breastworks. Dr. Schafer reported the defenses "consisted of a vertical-log stockade, the timbers sharpened at the top, anchored on the St. Johns River at Hogan's Creek and McCoy Creek. The stockade stretched in a half circle around the town and [had] several fortified redoubts. A twelve-foot-deep moat was dug around the stockade's outer walls to further strengthen the defenses."[35]

A writer of the Fifty-fifth Massachusetts Volunteer Regiment wrote,

---

[33] Ibid.; "From the Coast," *Augusta [GA] Daily Constitutionalist*, 15 February 1864, 1.

[34] "A Successful Raid Up the St. Mary's"; "1843. Ernestine (Strong) Alberti to Lydia (Strong) Clapp."

[35] Schafer, *Thunder on the River*, 191.

> [S]ince the retreat [from Olustee] our men have been engaged in intrenching [entrenching] the town on three side[s] of it. [The troops] are now protected by breastworks with three or four batteries and redoubts. Three of the field works are garrisoned by colored troops: Redoubt Fribley by a detachment of the 55th Massachusetts Volunteers, and Redoubts Reed and Sammons by detachments of the 3rd US Colored Troops. These batteries are erected on little eminences around town, and command the face of the country (which is remarkably level) for some distance in all directions.[36]

An attitude of bitterness and discontent seized the Union Army in North Florida after the Olustee fight. A correspondent from Jacksonville reported, "Our officers and men place but little confidence in the commanding General [Seymour], and among the troops of all arms can be heard expressions of distrust regarding him. He has never been successful and has, in different instances, been the means of having men uselessly slaughtered." As a result, a number of the bluecoats began deserting to the Rebels. A Georgian, temporarily stationed in Florida, averred, "About two weeks ago, the Yankee deserters…[began coming] into our lines in squads from two to five a day. I cannot tell how many came in, but there is quite a large number, and they brought with them their fine government horses." The turncoats loudly proclaimed their disgust at the idea of dying for Black people (though that may have been an example of saying what they believed the graycoats wanted to hear).[37]

The USCT had actually performed quite well in the battle, but after returning to Jacksonville, the Negro soldiers chose that inopportune time to address an inequity that gnawed at their souls. The Black troops received four dollars less a month than White troops of the same rank. Sgt. William Walker, of the Third USCT, was killed by a firing squad for persuading his men "to stack arms and refuse to do their duty" until they received wages equal to White soldiers. One of the Negro soldiers wrote to a New York paper, "[W]e are living a tyranny [as] inexcusable as slavery itself." The governor of Massachusetts weighed in on the controversy, averring that the blue

---

[36] Ibid; "From the Fifty-fifth Massachusetts Regiment," *[New York, NY] Weekly Anglo-African*, 23 April 1864, 2–3.

[37] "From the Fifty-fifth Massachusetts Regiment," 23 April 1864, 2–3; "From Florida," *Augusta [GA] Daily Constitutionalist*, 29 April 1864.

uniform gave Black soldiers dignity, "[but] the [US] government means to disgrace and degrade him [Black soldiers], so that he may be [seen as]...only a nigger." It took more than a year for the US Congress to address the inequity; but on 3 March 1865, Congress finally passed a law granting full pay to the Black troops.[38]

The Confederate command situation also remained in a state of flux. Gen. Gardner, who seemingly hoped to be named to replace Finegan, sent a series of reports to Beauregard regarding the poor state of affairs in Northeast Florida. Maj. Gen. James Patton Anderson, a veteran commander, soon sent Gardner back to his military backwater and retained the Irishman to lead the Rebel troops around Jacksonville. "Old Barney's" claims of sending "repeated orders to Colonel Smith, commanding the cavalry on his flanks and to continue in pursuit," were apparently taken at face value.[39]

THE ST. JOHNS RIVER

In the days following the Battle of Olustee, both Rebel and Yankee reinforcements poured into Northeast Florida. Gen. P. G. T. Beauregard even journeyed from Charleston to briefly oversee operations Florida. All of the Confederate leaders seemed to realize that an all-out onslaught on Jacksonville would be tantamount to suicide. As Gen. Patton Anderson observed, such a battle would cause a "great sacrifice of life, and to no purpose since his [Union] gunboats would prevent us from holding it." Instead, the Southerners began constructing fortifications along McGirts Creek, twelve miles from the River City. Conscripted slaves from as far away as Marion County were sent to help with digging the entrenchments and the construction of breastworks.[40]

The area between the two armies became a "no man's land," patrolled almost constantly by both Confederate and Union horsemen. During the last week of February and the first week of March, a series of skirmishes occurred near Cedar Creek as the two groups probed for weaknesses. Most of those actions turned out to be bloodless encounters between Col. Guy Henry's

---

[38] "From the Fifty-fifth Massachusetts Regiment," 23 April 1864, 2–3; Beard, "$10 a Month."

[39] Schafer, *Thunder on the River*, 198–200; Gleeson, *Erin Go Gray*, 17–18; *ORA*, ser. 1, vol. 35, pt. 1, 332.

[40] Coles, "Far from Fields of Glory," 196–202; *ORA*, ser. 1, vol. 35, pt. 1, 334–38.

mounted infantry and the 2nd Florida Cavalry. A Rebel trooper described a typical encounter. "On Saturday," he wrote, "the enemy, both cavalry and infantry, advanced [on] our line of pickets. What loss (if any) cannot be ascertained." Two days later, "a courier came in and reported the Yankees were across Cedar Creek. We were soon mounted and ready to make the attack, and before we could get in line the Yankees retired."[41]

Around 1 March, the encounters took a more serious turn. Col. Robert Anderson led two Rebel units—one of cavalry and Savannah's famed Chatham Artillery—toward Cedar Creek. This time the wily Col. Henry had a surprise set; as the graycoats crossed the stream, he sprang his trap. A Rebel wrote,

> About 10 o'clock skirmishing began, [with] the enemy falling back. As our cavalry reached Cedar Creek, they were...ambushed [by a] party of Yankees with two pieces of artillery. The bridge across this creek being broken in, our troops were exposed to their fire while it was being repaired. The Chatham Artillery replied vigorously, driving the enemy from their position. The skirmishing continued until 3 p.m., when the Yankees retired upon Jacksonville...We lost seven men killed and twenty-two wounded.[42]

In the midst of that fighting, as Capt. Winston Stephens had turned to give orders to his unit, a Union sharpshooter shot him in the back. Winston's brother, Swepson, wrote to his sister-in-law, Olivia, "He [Winston] fell before I could reach him. I dismounted and took him up and [placed] him on my horse and got up behind him and took him out in that way." The Yankees had long observed Capt. Stephens's reckless courage. Charles Currier, of the 40th Massachusetts, wrote,

> Throughout the entire campaign, whenever a force of the enemy were encountered, there had always been seen among them, an officer who rode a cream colored horse. He was a gallant fellow, and much ammunition had been expended in the attempt to unseat him.... During the action of March 1...one of them [Currier's comrades] succeeded in bringing down the rider of the cream

---

[41] Schafer, *Thunder on the River*, 202–203; "From Florida," *Charleston [SC] Mercury*, 12 April 1864, 2.

[42] "From Florida"; "Further from Florida," *Augusta [GA] Daily Constitutionalist*, 8 March 1864, 1.

> colored horse: the shock of his death was so great, as to cause a temporary [cessation] of their [the Rebel] fire, whereupon our men set up a great cheer.[43]

For Winston's wife, who had taken refuge at Thomasville, Georgia, it proved to be the beginning of a horrendous week. Two days after the news reached her of Winston's death, Olivia's mother died, and Olivia gave birth to a premature male child. She named him Winston to honor her husband. Soon thereafter, her young daughter became very ill but eventually recovered. In her diary she wrote, "I now begin as it were a new life and I pray the Lord will give me strength to bear up under this great affliction, and with His help and [the] examples of those two dear ones now with him I may be enabled to do my duty in this life and be prepared when the Lord calls to meet them in that 'better world' where there will be no more parting and no more sorrow."[44]

The Union situation in North Florida worsened through March and April 1864. In early March, Gen. Truman Seymour received orders transferring him to Gen. U. S. Grant's Army of the Potomac. Brig. Gen. John P. Hatch replaced the "goat" of Olustee, and the Union leadership at Jacksonville became a sort of revolving door with seven commanders in one year. A modern historian observed, "These frequent command changes undoubtedly affected Union morale, not to mention their impact on a coherent, defined strategy within the state."[45]

PALATKA

The bluecoats retained control of Jacksonville, Amelia Island, and St. Augustine, but Gen. Hatch decided to extend their territory to include Palatka (which Northern newspapers, almost without exception, spelled as "Pilatka"). They apparently hoped that with freedom again as close as the river, this would encourage more runaway slaves and deserters to come into the Union camp. Many Federal leaders in Washington, and, to a lesser degree, in North Florida, still clung to the chimera that most White Floridians yearned to

[43] Schafer, *Thunder on the River*, 202–203; Coles, "Far from Fields of Glory," 189–91.

[44] Schafer, *Thunder on the River*, 204; Blakely, et al., *Rose Cottage Chronicles*, 328–29.

[45] Coles, "Far from Fields of Glory," 203–204.

return to the Union—"their first love." Many bluecoats realized what their leaders failed to acknowledge—that native Floridians were largely loyal Confederates. A reporter for the *New York Tribune* wrote,

> Deserters come into our lines nearly every day, but they are mostly persons of Northern birth, and were pressed into the Rebel service. Native Floridians, especially the wreckers [Crackers] and upper classes have not, as a general thing, any more idea of abandoning the Rebel cause than the leopard does of changing its spots. The important point of agreement [among Floridians] is the Rebels mean to hold Florida, let the cost be what it may.[46]

On the evening of 1 March, three US vessels left Jacksonville, bound for Palatka. The *Maple Leaf, General Hunter*, and *Charles B. Houghton* made the sixty-mile trip upriver after dark and without running lights to keep Confederate guerrillas from sniping at them from the riverbanks. The Federal commander at Palatka was Col. William B. Barton. His troops included soldiers from the Forty-seventh, Forty-eighth, and One Hundred Fifteenth New York Regiments, the Fifty-fifth Massachusetts (USCT), and Company C of the Third Rhode Island Artillery. The exact number of Federals garrisoned at Palatka is not known. Capt. J. J. Dickison believed the enemy had five thousand men, but the bluecoats at Palatka likely numbered from fifteen hundred to two thousand men. Initially, the Union soldiers quickly spread throughout the town, "commandeering" anything that caught their eye, but Col. Barton soon put them to work digging entrenchments and gun emplacements. The Federals felt complete confidence in their ability to hold their position against any force the Rebels could throw at them. A member of the Forty-eighth New York boasted, "I don't think they [the Confederates] are fool enough to come down here and get killed. We have got this place arranged so as to keep back a hundred thousand troops."[47]

The Union expedition to Palatka represented a major problem for Patton Anderson and the Florida Confederates. The new fortress might provide

---

[46] Ibid.; "From the Department of the South," *New York Tribune*, 1 April 1864, 1.

[47] Schafer, *Thunder on the River*, 204–205; Coles, "Far from Fields of Glory," 202–205; R. Martin, "Great River War on the St. Johns"; Evans, ed. (and J. J. Dickison), *Confederate Military History* (Florida), 11:85–86.

a haven for runaway slaves and deserters, but they could live with that loss. The greater

> concern was protecting the supply line of beef and other commissary supplies from Florida that had become so vital to Confederate armies. Recent setbacks on the Mississippi and Tennessee had resulted in drastic [food] shortages for the major Confederate armies in the East; [Rebel] officials feared the loss of Florida and its cattle herds would have catastrophic consequences. Regaining control of the St. Johns River was imperative if Florida was to remain in Confederate hands.[48]

The expedition to Palatka paid some immediate dividends for the Unionists. A few Confederate light vessels had evaded capture by slipping into smaller rivers and streams where larger Union gunships could not follow. While a minor irritant in the greater war effort, Col. Barton wanted to rid the waterways of these boats. "The expedition [to capture the vessels] was considered full of danger," but Acting Master John C. Champion, and his first mate, F. W. Sanborn, volunteered to lead the search, accompanied by a dozen US Marines. Within a few days, the raiding party had captured the *Sumter* and *Hattie A. Brock*, but they failed to locate the *Silver Springs*. One of the most touching—or comical—episodes of the war (according to the reader's perspective) occurred while Champion's weary raiding party returned to Palatka with the *Hattie A. Brock*. As the unit and their prize passed through the little town of Enterprise, Hattie Brock, the ship's namesake and an aging Southern belle, stood on her veranda to watch "her" ship go by. Seeing the vessel in Yankee hands, she was "eloquent in her grief, and apparently denounced the captors in language that no proper lady would employ." The writer added that even the hard-bitten Marines "were glad to be away" as soon as they could.[49]

Despite this initial setback, Patton Anderson had two weapons he believed would turn the tide along the St. Johns River. The first of these was an "infernal device" the Confederates called "torpedoes." That designation might seem odd to modern readers, as the Civil War "torpedo" worked very much like a modern naval mine. A rather shadowy figure, Capt. E. Pliny Bryan, seems to have been the "father" of the device, and Gen. Beauregard

---

[48] Schafer, *Thunder on the River*, 205.

[49] Ibid., 222; "From the Department of the South."

had been using him to set torpedoes in Charleston Harbor; "On March 11...Bryan...was ordered to begin torpedo operations on the St. Johns." Bryan and a small squad of volunteers commenced constructing and deploying the devices in the channel off Mandarin Point. Then, they sat back to see if the weapon would be successful.[50]

The Federals had done everything in their power to make their gunboats invulnerable to threats above the waterline. A Massachusetts newspaper described a typical vessel's protection:

> [L]ittle will be presented above the water line save the muzzles of the guns and the top of the smoke-stacks. All the machinery is under water, and this part is so divided into water-tight compartments, that, in case a [cannon]ball should enter one, it can do not material damage. The sides are sloped at an angle of 45 degrees, and are covered with iron plates two and one-half inches in thickness, securely bolted to each other and through the heavy timber beneath. Each plate has been submitted to adequate tests, and calculated to resist any missile known to modern warfare.... The magazine is perfectly inaccessible to accidents of any kind, and placed below the water line.

In the years before submarine warfare, the Federals had done everything in their power to ensure the safety of their sailors.[51]

On the dark morning of 1 April, the *Maple Leaf* left Palatka with sixty soldiers and civilians aboard, and a crew of at least five. The six-hundred-ton ship, leased from a Canadian company, had a somewhat checkered history. The previous year, seventy Confederate prisoners, being transported to a Northern prison, had rushed their guards, disarmed them, and rowed the ship's launches to freedom in North Carolina. Fifteen miles south of Jacksonville, the *Maple Leaf* "ran across a torpedo, which exploded under her about thirty feet from her bow, and caused her to sink in a few minutes." The Northern press expressed outrage at the cowardly action. A Philadelphia correspondent scribbled, "Rebel chivalry boasts of numerous infernal and

[50] Coles and Waters, "Indian Fighter," 52–55; R. Martin, "Great River War on the St. Johns."

[51] *Worchester [MA] Aegis and Transcript*, 7 December 1861, quoted in Koblas, *Swamp Fox*, 86.

barbaric acts, and this blowing up of steamers, by torpedoes, must be classed among them."[52]

The Federals, of course, set out to make life difficult for the men who placed the "infernal devices" in the river—and those who sheltered them. Gen. Hatch sent Col. Guy Henry and the Fortieth Massachusetts, who had so successfully spearheaded the initial action in the Olustee Campaign, to "destroy [all] the houses along the St. Johns River." This represented an attempt to remove the citizens who gave refuge to Pliny Bryan, Jacob Mickler, and the torpedo squads. Henry diligently carried out his orders—except in one instance. Henry left the Fleming family's plantation, "Hibernia," standing. The Flemings were avid secessionists, and two of their sons (Charles and Francis) had been among the first Floridians sent to the Rebel army in Virginia. Col. Henry, however, "had been a guest of the Flemings before the war and [was] a cousin of Charles and Francis Fleming." In another effort to deny aid and comfort to the torpedo squad, the Yankees also landed troops at Green Cove Springs, in Clay County. They took Capt. Harry Henderson and George Clinch prisoner and ordered suspected Rebels to move "beyond the lines." A Southern newspaper lamented that the refugees had been "driven into the woods without shelter or provisions." The bluecoats also had orders to arrest Paul Arnau, the former mayor of St. Augustine who had formed the "Coast Guard" unit that had darkened the lighthouses along Florida's Atlantic coast. Arnau, however, escaped the Union raiders, having previously taken residence at Lake City.[53]

The Federals cleared all the "infernal devices" they could find from the area around Mandarin, but that night the Rebels simply put more into the St. Johns. On 16 April, the *General Hunter* joined the *Maple Leaf* on the bottom of the river. The explosion that sank the 470-ton vessel "lifted a mountain of water and the... *General Hunter* from the water." The ship sank rapidly, but only one man lost his life. The rest of the passengers and crew "barely managed to escape." A few weeks later the *Harriet Weed*, a smaller vessel, "was literally blown to atoms." In July, the *Alice Price* became the last

[52] R. Martin, "Great River War on the St. Johns"; "Affairs in Florida," *Philadelphia Inquirer*, 14 April 1864, 1.

[53] "The Florida Columbian," *Columbus [GA] Daily Enquirer*, 2 July 1864, 2; "Calendar of Civil War Activity 1862–1864."

known victim of the "infernal devices" in the St. Johns. Palatka, by that time, had been abandoned for three months.[54]

On 12 April, the Yankees evacuated their position at Palatka, citing the difficulties encountered in supplying the outpost. A member of the Fifty-fifth Massachusetts (USCT) summed up the feelings of many of the common soldiers:

> The Fruits of this expensive little expedition were two miniature, worn-out steamers [the *Sumter* and *Hattie A. Brock*]...with their crews and a few dollars worth of cotton and turpentine, a few horses, and about twenty-five contrabands. These acquisitions cost our government the loss of two of our finest transports, the *Maple Leaf* first, and then the *General Hunter*, said to be the swiftest steamer of her kind, were blown to atoms by torpedoes, placed in the river by the barbarous rebels.... The aggregate cost of those two steamers was about $130,000. So much for the Pitlaka occupation.[55]

The second "weapon" Gen. Anderson possessed, in his efforts to drive the bluecoats out of Palatka, were his guerrillas. Col. McCormick took part of the 2nd Florida Cavalry, Capt. M. S. G. Gary's Marion County horsemen, and a few torpedo-men, to the area around Broward's Neck and Yellow Bluff to keep an eye on the Union forces northeast of Jacksonville. Anderson stationed Gen. Finegan, with a large contingent of infantry, artillery, and cavalry, to block any direct move into the middle of the state. To deal with the Union soldiers at Palatka, Anderson sent Capt. J. J. Dickison, reinforced by a contingent from Lt. Col. John L. Harris's 4th Georgia Cavalry, to keep a watch on the Federal outpost in Putnam County.[56]

For the first week or two the situation at Palatka settled into a predictable pattern. A Northern correspondent wrote, "Our outposts at Palatka seems to excite the belligerency of the Confederates, and consequently they make a daily and sometimes semi-diurnal [dusk and dawn] attacks on our pickets...Almost every day a company or two of Rebel cavalry dash in and fire

---

[54] R. Martin, "Great River War on the St. Johns."

[55] "From the Fifty-fifth Regiment Massachusetts Volunteers," *New York [NY] Weekly Anglo-African*, 21 May 1864, 2.

[56] Evans, ed. (and J. J. Dickison), *Confederate Military History* (Florida), 11:85–86.

upon our pickets, and in return get shelled by our batteries." Another Union soldier described a typical encounter:

> The cavalry of the enemy...came down upon our mounted pickets...using a couple of houses on the hill as screens behind which to fire upon our men. Another force came against our pickets on the left, who were pressed for a few moments.... The *Ottawa*, lying off the town, fired four shots from her two hundred pounder [cannon] and ten from her thirty pounder [cannon] during the skirmish. Strange to say, no one was hurt on our side, and the Rebels...must have been equally secure from injury.

Even though greatly outnumbered, Dickison found that situation unacceptable. Never one to hide his light under a bushel, the Rebel partisan wrote, "A detachment of 16 men under Captain Dickison...was met by a superior force of the enemy and after a hot skirmish which lasted forty minutes, holding their position without giving an inch, the enemy was reinforced and our men fell back in good order without loss. The enemy loss was 5 killed and 8 wounded." On another occasion, the Rebel guerrilla leader sent a small contingent, led by his second-in-command, Lt. W. J. McEaddy, who captured eight of the enemy pickets. A few days later, Dickison, and Co. H, fought two Federal regiments for four hours, reportedly killing eleven and capturing twenty-two of the bluecoats.[57]

With the Rebels thus occupied, the Federals from St. Augustine made a raid south into Volusia County, reportedly capturing fifteen hundred head of cattle. The Union forces quickly returned to the Ancient City, driving their meat on the hoof to "a grazing area near Pellicer Creek south of St. Augustine." They also spread the word to area residents that Uncle Sam would pay five dollars a head for each cow they delivered to the Federals. (With Rebel script barely worth the paper it was printed on, this allowed residents of the hardscrabble region a chance to make some real money.) Some Floridians cooperated with the Yankees, like "Old [Tom] Petersen" while the Federals placed die-hard Rebel Moses Barber on a "captured dead or alive" list.[58]

---

[57] "Affairs in Florida," *Philadelphia Inquirer*, 14 April 1864,1; "A Dash on the Pickets," *New York Herald*, 24 March 1864, 1; Evans, ed. (and J. J. Dickison), *Confederate Military History* (Florida), 11:86.

[58] Schafer, *Thunder on the River*, 223. See also, Waters and Edmonds, *Small but Spartan Band*, 187.

On 14 April, Gen. Hatch ordered the evacuation of Palatka. The Federal soldiers departed happily, but the Union had no intention of abandoning the St. Johns to the Rebels. They established a new post at Picolata, thirty-five miles south of Jacksonville on the eastern bank of the St. Johns, and at Orange Mills. Reportedly, all that remained of that hamlet was two or three houses, some entrenchments, and a wharf. A South Carolina newspaper used the fortifications there as an excuse for a heavy-handed attempt at sarcasm: "They [the bluecoats] are fortifying Orange Mills, ten miles north of Palatka. Their object seems to be to plunder and steal cattle and hogs—which at this time, undoubtedly pays better than fighting."[59]

GEN. BIRNEY'S RAIDS

Brig. Gen. William Birney assumed command of US forces at Jacksonville in mid-1864. He immediately set in motion a series of operations designed to strike a blow at the enemy, while also strengthening the Union cause in East Florida. Birney had been born in Huntsville, Alabama, but his father, a Southern Abolitionist, had moved to the North before the war in order to escape the persecution he and his family received due to their unpopular belief. Birney had been tried in the fire of battle at First Manassas, Fredericksburg, and Chancellorsville, and, as might be expected, strongly encouraged the enlistment of Black soldiers wherever he served.[60]

Almost immediately after his arrival at Jacksonville, the aggressive Birney launched his first of several raids. He initially sent a detachment of mounted bluecoats toward Trout Creek (now Trout River). A small party of Confederate cavalry, likely Lt. Peter Cone and about twenty-five members of Co. K, 2nd Florida Cavalry, briefly contested the crossing of the creek but quickly retreated under the weight of the vastly superior Yankee firepower. Once across the Trout, Birney divided his men, sending "one squad of cavalry due west toward the [Little] St. Mary's, and another [northward] to Callahan, in Nassau County, on the [Florida] railroad." The first column of Federal raiders managed to destroy two rail cars, loaded with iron, as well as a telegraph office, and a small cache of weapons and ammunition. They also brought off "a dozen negroes."[61]

---

[59] Schafer, *Thunder on the River*, 210; 222–24; "Yankee Movements in Florida," *Charleston [SC] Mercury*, 20 May 1864, 2.

[60] "General William Birney, USA"; Warner, *Generals in Blue*, 35.

[61] *ORA*, ser. 1, vol. 35, pt. 1, 410–13; Schafer, *Thunder on the River*, 228–30.

The expedition toward the St. Marys, however, was a diversionary operation designed to keep the Rebels occupied while Birney's second column accomplished the raid's primary goal. According to the Union commander, "This advance covered the operations of a party sent up the Nassau River to Holmes' Mill, for the purpose of taking it [the mill] away. The mill is one of the first ever erected in Florida, containing gang, rotary, and circular saw machinery, and worth now probably $50,000 when in operation." Birney averred that the mill could "turn out from 40,000 to 50,000 feet of lumber daily." Birney's plan called for bringing the mill machinery to Jacksonville, by ship. He explained, "I propose having it erected at once on the former site of the Empire Mills, about two miles below Jacksonville, on the east side of the Saint Johns [River]." The Federal War Machine's appetite for processed lumber continued to be almost insatiable. The bluecoats had conducted a raid at New Smyrna, in 1861, to gather lumber, and had gone into Nassau County, at least twice, for the same purpose. The captured mill would alleviate the need for raids into Confederate territory, as it would allow the bluecoats to cut and process their own lumber at the River City.[62]

While his troops were north of the St. Johns, Gen. Birney took the opportunity to strike a blow at Confederate partisans in Duval County—particularly the Broward family. The Broward clan, a wealthy and politically influential band, had been among Florida's earliest advocates of secession. In 1860, Miss Helen Broward, and several of her female friends, had presented Gov. Madison Starke Perry, with a "secession flag sewn by 'the ladies of Broward Neck.'" Many of the Broward children and their close kin had enlisted in the Confederate army. Those who remained at home carried on an independent guerrilla war with the Federal gunboats and had reportedly furnished aid and comfort to the men who put the torpedoes in the St. John's River. The bluecoats reported, "[T]he Broward clan had come to be regarded as participants in the most damaging guerrilla actions in East Florida." They particularly wanted to kill, or capture, nineteen-year-old Washington Broward, who had engaged in a personal vendetta against Union soldiers and deserters from the Rebel armies. He and his family also provided intelligence and supplies to Confederate scouts stationed on the outskirts of Jacksonville.[63]

---

[62] *ORA*, ser. 1, vol. 35, pt. 1, 410.

[63] Broward, *Broward Family*, 63–88; Schafer, *Thunder on the River*, 227–28.

To conduct their campaign against the Broward clan, Gen. Birney enlisted the aid of the Tison family. The Browards regarded the Tisons as "a pack of cutthroat murderers," totally devoid of "honor or truth," that had used the Civil War as a "free license for murder and theft." While the truth of these charges may never be known, there seems to be a nugget of truth regarding this Snopesian brood. Several of the Tisons not only deserted from the Confederate army but also reportedly killed captive Rebels in cold blood. In one year they went from being poor, landless tenants to some of the largest cowmen in Duval County.[64]

Lt. Col. McCormick could muster fewer than one hundred Rebel cavalrymen to oppose the expedition to the sawmill, and, aided by deserters and pro-Union citizens, the Federals were able to capture several scouts who had long been a thorn in their side. The Tison family took two as prisoner 2nd Cavalry scouts, Ashmun Houston and Washington Turner, and Col. McCormick reported, "We afterwards found they [the scouts] had been brutally murdered." The Broward family archives insisted, "[The Tisons] feasting in Satanic glee upon [these] horrors...fell upon them [Turner and Houston], wounded and bound, and in cold blood, cut their throats." The Union cavalry, likely abetted by deserters, also "arrested Joseph Hagans and Washington Broward...and carried off Mr. Geiger's negroes, and burnt the home of Joel Wingate," a well-known Rebel scout. According to Dr. Schafer, "Washington Broward was brought to Jacksonville in a fashion calculated to embarrass and infuriate; shackled and led by black troops [and] in a column that included liberated slaves." Washington's father, Napoleon Broward, wrote to Gov. Milton, pleading with him to intercede with Federal authorities for the release of Washington Broward, Ansel Rouse, James Turner, W. A. Webb, and Joseph Hagans. If Florida's chief executive tried, he failed to obtain their freedom. The Yankees sent young Broward to a prison at Hilton Head, South Carolina. He died there a few months later while still a teenager.[65]

---

[64] Broward, *Broward Family*, 123–31. Reprint of John Holn, *Tison Exposa.* Holn was the Broward family attorney, and according to Broward family history, the Tison family succeeded in buying, and destroying, all but two or three copies of Holn's incendiary pamphlet.

[65] Broward, *Broward Family*, 85–88; Schafer, *Thunder on the River*, 227–28; *ORA*, ser. 1, vol. 35, pt. 1, 411–13; Holn, *Tison Exposa.* Broward's youngest brother would be elected governor of Florida in the early 1900s.

About a week after their return from the first raid, Birney left Jacksonville for an incursion into Volusia County. He took six companies of the Thirty-fifth USCT (mounted) to Palatka, transporting them aboard the *Mary Benton*. Three companies of the One Hundred Fifty-seventh New York, stationed at Picolata, soon followed on the transport ship *Harriet A. Weed*. The ships proceeded to Welaka, where Northern newspapers reported, "the Rebel pickets were seen to run precipitately." Employing Birney's favorite tactic, Col. Noble, from St. Augustine, with several companies from the Seventeenth Connecticut and Seventy-fifth Ohio, marched south to aid Birney's raid. In all, the Union contingent likely numbered at least one thousand men.[66]

Historians disagree regarding the reason for Birney's incursion into the Volusia County backcountry. Jeff Grzelak suggested the raid constituted a "last attempt of Union forces to redeem themselves for the February defeat at Olustee." David Coles postulated that the genesis of the foray came when "a refugee [who] reported to Birney...that about 400 horsemen under J. J. Dickison were in the vicinity of Fort Gates...building rafts or 'lighters' on which to cross the river and raid the Union controlled east bank [of the St. Johns River]." The report also claimed that seventeen Union sympathizers had been captured by Rebels "and hung on the spot." The report was either pure fiction or based upon erroneous intelligence. (Dickison never commanded a unit of four hundred men in his entire military career, and, unlike many guerrillas, he rarely, if ever, killed captives—Black or White.) Possibly both motives factored into Birney's decision to raid into Volusia County, but "confiscating cattle, horses, edibles, and disrupting Confederate supply lines" likely constituted the predominate reasons for the incursion.[67]

The Yankee raiders, meeting almost no opposition, rapidly moved through western Volusia County. As they swept through the sparsely settled region, Birney left small Union outposts, or garrisons, at Welaka, Saunders, and Volusia. His bluecoats were successful in gathering large quantities of beef, mules, and other assets to be "retained for the United States Quartermaster's Department." As the Federals continued their advance, numerous

[66] "A Great Raid by General Birney in Florida," *Sacramento Daily Union*, 17 June 1864, 2; Williamson, "Birney's Raid in the News"; Coles, "Volusia County," 44–45.

[67] Coles, "Volusia County," 44–45; Williamson, "Birney's Raid in the News"; "A Great Raid by General Birney in Florida."

deserters and conscript evaders emerged from the swamps and sloughs where they had taken refuge. They pointed out the location of hidden caches of supplies and identified Rebel supporters. At De Leon Springs, the Federals also destroyed John Starke's Spring Garden Plantation. Starke's three gins and his water-powered gristmill had furnished a large quantity of cornmeal to East Florida Confederates. From there, the raiders turned toward the Atlantic coast.[68]

Not surprisingly, the Yankees wanted to check the situation at New Smyrna. The village may have been destroyed the previous year by the ships of the US Blockading Squadron, but blockade-runners still regularly used Mosquito Inlet as a safe haven before or after making the run to the Bahamas. According to a newspaper account,

> "The forces [Birney's], having combined, penetrated as far a [New] Smyrna, on the east coast of Florida, and seized two schooners laden with cotton, prepared to run the blockade. Here, also was a large amount of cotton stored for shipment…. The whole force, without the loss or hurt of a man, regained Jacksonville on the sixth of May. The Union forces captured 5,000 or 6,000 cattle, a large number of horses, mules, and other means of transportation, and more than a million dollars worth of cotton."

A history of the Seventy-fifth Ohio adds more detail concerning the expedition:

> Gen. Birney…sent Colonel [A. L.] Harris with the 75th on a raid to the headwaters of the St. Johns and Kissinnee [Kissimmee] Rivers, for the purpose of breaking up a system of blockade-running carried on in the Rebel cause along the Indian River. The regiment proceeded as far as Lake Harmer [Harney] when it [the Seventy-fifth Ohio] divided in two parts; one division was sent to [New] Smyrna…where it captured two schooners loaded with cotton, and sent them, together with their crew safely to St. Augustine. The remainder of the regiment continued their march southward, captured and destroyed five hundred bales of cotton, destroyed three salt furnaces, and burned a large lot of resin, tar, and turpentine stored by the Rebels at Sand Point [modern Titusville] on the

[68] Ibid.

> Indian River, from which point blockade-runners...carried it to Nassau.... From this point the detachment made its way to the headwaters of the Kissinnee [Kissimmee] River, and captured a large lot of cattle driven there by Rebel owners to prevent them from falling into the hands of the National [U.S.] army.

Despite the seemingly overwhelming success of the incursion, Gen. Birney blundered badly by leaving three small outposts in the isolated region. That miscalculation would cost the Federals dearly in the days to come.[69]

On the night of 20 May, Capt. Dickison and fifty of his men crossed the St. Johns at a narrow point of the waterway on three decrepit rafts.

> Approaching [the Welaka garrison] at night...[the Rebels] surprised the Federals, who seem to have not taken the proper defensive precautions. The Yankees had used a building near the river as a barracks. Nearby three roads joined together, but guards had not been posted on the northern route...The Welaka garrison surrendered quickly to Dickison's troops, without a shot being fired in their defense.

Another source reported, "The captain and a portion of Co. B, 17th Connecticut Infantry, meekly laid down their guns. Estimates as to the number captured vary, but probably about thirty-five men became Dickison's prisoners at Welaka.[70]

Dickison and his men returned to the St. Johns River with his captives. After a brief rest, the "Swamp Fox" informed his men that he also intended to capture the small enemy encampment at Fort Butler. (Modern and contemporary sources vary as to the name of the outpost. Some refer to the picket post as Saunders. Dickison referred to it as "Fort Butler," while many 1864 Union reports called it "Fort Gates.") With a small force—about twenty-five men—the Rebels bluffed the Yankees into surrendering. Dickison reported, "Captain Dickison then made a rapid advance with his detachment on the enemy post, 2 miles distant, the location being shown by a bright campfire.

---

[69] "A Great Raid by General Birney in Florida"; Reid, *Ohio in the War*, 2:436–38.

[70] Coles, "Volusia County," 44–47; Weinert, "Swamp Fox of Florida," 6–7; "Confederate Swamp Fox." The captured Federals blamed members of the Winston Stephens's family for alerting Dickison of the unguarded trail while Dickison praised Lt. McEaddy, of his unit, for the discovery.

Moving cautiously within two hundred yards Lieutenant McEaddy was sent forward a demand to surrender. The [Federal] captain in command held a short parley, and very reluctantly complied. On the return trip they took in hand 12 slaves [probably contrabands making their way to Union-controlled territory]." The Florida guerrilla leader reported that he captured a total of eighty-eight infantry and six cavalrymen.[71]

The news of the raids threw the Union command in Northeast Florida into an uproar. One Yankee outpost, at Volusia, yet remained in the region far removed from Jacksonville and St. Augustine. Col. William Noble, commanding at the Ancient City, had personally gone to the fortress to oversee making it "invulnerable" to guerrilla attacks. He spent almost two weeks and had the bluecoats "put the camp in an open field, where they could not be surprised, and [they] had constructed a very handsome and effective redoubt." Despite Noble's hard work, Gen. Birney decided to evacuate the men from the lonely, isolated outpost. To extricate the bluecoats he sent the gunboat *Columbine*.[72]

The first rule of guerrilla warfare has always been "Never attack where the enemy anticipates—attack where least expected." The Confederates had, for most of the conflict, been content to snipe at the Yankee ships from wide banks of the upper St. Johns. Two units, however, had recently joined Dickison. Capt. H. A. Gray, of Co. B, 2nd Florida Cavalry, who had taken com mand of that unit after the death of Winston Stephens, arrived with about twenty-five men. He was followed soon thereafter by Lt. Mortimer Bates, with two guns from the Milton Light Artillery. Within minutes, a courier arrived at Dickison's camp shouting, "The river is full of Yankee gunboats coming up." Dickison quickly moved his unit to Horse Landing, where the banks of the St. Johns narrowed considerably.[73]

Gen. George H. Gordon led the naval expedition to evacuate the Union detachment at Volusia. His troops boarded the gunboat *Ottawa* and armed steam tug *Columbine*, and Gordon also sent a relief column, consisting of six or seven hundred men from the Thirty-fifth USCT and One Hundred fifty-

---

[71] Ibid.; Evans, ed. (and J. J. Dickison), *Confederate Military History* (Florida), 11:88–89; Koblas, *Swamp Fox*, 81–83.

[72] Ibid.; Letter from Col. Noble," *Stamford [CT] Advocate*, 24 June 1864, 2.

[73] Evans, ed. (and J. J. Dickison), *Confederate Military History* (Florida), 11:88–90; Koblas, *Swamp Fox*, 86–87; David Cook, "Confederate hero captures gunboat," *Ocala Star-Banner*, 3 November 1991, 13C.

seventh New York, overland, through the East Florida woodlands. The shallow depth of the water at Horse Landing prevented the *Ottawa* from moving through the narrows, but the *Columbine*, with its lighter draft, passed upriver. As the vessel began the return trip, Acting Ensign Frank Sanborn ordered his bluecoats "to quarters," expecting Rebel snipers would take a few potshots as the vessel passed Horse Landing.[74]

A member of the Thirty-fifth USCT provided an excellent account of the ensuing engagement. "When we arrived at Palatka," he wrote,

> between 45 and 55 men went on board the gunboat *Columbine*, and in company with the gunboat *Ottawa*, went up the river. The object...was to bring away some of the 17th Connecticut Volunteers, who were stationed at Volusia. The *Columbine* went as far [south] as Lake George, when the lowness of the water prevented her going further. A small boat, however, went to Volusia and ordered the detachment there to proceed overland to St. Augustine.... [On the return trip, as the *Columbine*] approached Horse Landing, where the river is quite narrow, they [the *Columbine*'s sailors] shelled the wood, as they did when they were ascending, but when directly opposite [the Landing] the rebels opened with grape and canister from a battery they had placed there. Our party returned the fire as well as they could, but a [Rebel] shot cut the tiller rope, leaving the boat unmanageable, and it went aground with the stern [pointing] directly toward the rebels, so their batteries could sweep the entire deck.... The captain of the boat, finding it useless to attempt to defend the boat, surrendered; five of our men and two of the crew escaped by jumping overboard and swimming ashore; the rest were taken prisoner or drowned in attempting to escape.[75]

Dickison reported that Bates and his gunners "continued to pour canister and solid shot, while our sharpshooters kept a constant and well-directed fire until she [*Columbine*] became unable to manage her guns. Our battery shot with much precision...riddling her badly." The riflemen also made life miserable for those who remained on the disabled *Columbine*. The guerrilla

---

[74] Weinert, "Swamp Fox of Florida," 8–9.

[75] "Letter from Florida," *Worchester [MA] National Aegis*, 18 July 1864 (written 31 May 1864), 1.

leader wrote, "Never did a command fight with more gallantry than the artillery and sharpshooters, every man displaying remarkable coolness and bravery." Sanborn withstood the barrage for almost forty-five minutes before raising the white flag. He rowed across the river in a launch and surrendered after receiving Dickison's pledge that the Unionists—Black and White—would be treated well. Although there are discrepancies the in various casualty reports, "Dickison put Federal losses at about twenty-five killed and drowned. Sixty-four Yankees fell into Rebel hands." In addition, the Florida horsemen salvaged a considerable quantity of weapons and other military hardware.[76]

News of Dickison's coup quickly spread throughout the state and Confederacy, giving Southern supporters a brief boost in morale. The losses embarrassed Union leaders in Northeast Florida but did little to change the outcome of the conflict. There was, however, no rest for the weary. Col. McCormick immediately dispatched Dickison and his men to patrol the area in Clay County around Green Cove Springs and Camp Bayard. A Dickison biographer reported, "During the month of June and part of July his [Dickison's] command at the outposts on the St. Johns River continued to perform effective service, frequently engaging in skirmishes with detachments of the enemy and capturing their pickets." The arrival of a detachment from Co. H, of the Second Florida Cavalry, allowed them to cover more territory and respond speedily to Union raids from Jacksonville and Palatka.[77]

FLORIDA TROOPS SENT TO VIRGINIA

By May 1864, the Southern Confederacy seemed to be plummeting toward extinction. Gen. Joseph E. Johnston, commanding the Rebel Army of Tennessee continued his retreat toward Atlanta after being maneuvered out of strong positions at Dalton and Resaca, Georgia. He would soon be forced south of the Etowah River, which Gen. W. T. Sherman had called "the Rubicon of Georgia." In Virginia, the slaughter at the Wilderness and Spotsylvania Court House was bleeding Lee's superb Army of Northern Virginia to death. Days before, Maj. Gen. John C. Breckinridge had been forced to use the young cadets at the Virginia Military Institute in order to win the Battle

[76] Coles, "Far from Fields of Glory," 216–17; *ORA*, ser. 1, vol. 35, pt. 1, 397; M. E. Dickison, *Dickison and His Men* 67.

[77] Coles, "Far from Fields of Glory," 217; M. E. Dickison, *Dickison and His Men* 75.

of New Market; in effect, "grinding the seed corn of the Confederacy," just to keep the enemy at bay.[78]

Once more, Richmond called on Florida for fighting men. On 16 May 1864, Maj. Gen. Patton Anderson received orders to send "one good brigade of infantry" to the Army of Northern Virginia. This meant that Florida would be stripped of virtually all of its fighting men to meet the crisis in Virginia. "Only two cavalry battalions, three artillery companies [and a few assorted guerrilla and home guard units] would remain to protect the state and safeguard the vital flow of beef and other food supplies to the Confederate military." The commander of the Department of Florida complied but warned the Davis government that the "order would likely lead to mass desertions among the new troops."[79]

Approximately twelve hundred men made the journey to the Old Dominion. These troops included the 6th Florida Battalion, commanded by Col. John M. Martin; the 1st Florida (Special) Battalion, directed by Col. Charles F. Hopkins; the 2nd Florida Battalion (formerly Partisan Rangers), led by Col. T. W. Brevard; and Col. James F. McClellan's 4th Battalion. Among the "Flowers" who went to Virginia in May 1864 were several men who had contributed greatly to the state's guerrilla operations. Cols. Brevard and Hopkins had seen action at St. Johns Bluff; Maj. Westcott at Tampa; and Capts. John W. Pearson, Sam Hope, S. M. G. Gary, and Samuel Mays had all previously participated in successful irregular actions. Brig. Gen. Joseph Finegan, the hero of Olustee, assumed command of the units sent to Lee's Army.[80]

Birney's Black Creek Raid

The Federals' last Northeast Florida operation of July 1864 vividly revealed the weakness of the Florida forces confronting the bluecoats at Jacksonville. (Most of J. J. Dickison's men had moved south from Clay County to counter a reported threat in the Palatka area.) Although no longer the commander at the River City, Gen. William Birney directed the operation, using a task force made up of both infantry and cavalry. His foot soldiers consisted of men from the Thirty-fifth USCT, while his horsemen came from the Fourth Massachusetts Mounted infantry and the Seventy-fifth Ohio Cavalry. Bluecoats likely

---

[78] Mosocco, *Chronological Tracking of the American Civil War*, 230–35.

[79] Waters, "Tell Them I Died Like a Confederate Soldier," 156–59.

[80] Waters and Edmonds, *Small but Spartan Band*, 120–24.

numbered about six hundred. Using his favorite tactic, Birney sent a separate group toward Nassau County, which included troops from the Seventh and Eighth USCT and, from St. Augustine, a portion of the Seventeenth Connecticut.[81]

Birney's contingent left Jacksonville and disembarked at Magnolia, in Clay County. At 1 a.m., the bluecoats left camp and proceeded toward the South Prong of Black Creek. The Union soldiers soon discovered that stream was too deep to ford and spent precious time building a wobbly bridge out of nearby fence rails. Despite being in an ideal position to contest, and possibly halt, the advance, Birney reported the Rebels did not mount a resistance. McCormick disputed that account, claiming Maj. Scott, with ninety-eight men, "met and repulsed from 300 to 500 of their infantry, driving them back across Black Creek." If so, Scott's men then chose to fall "back about five miles to a creek in his rear." Near Trail's Ridge, Birney's men again encountered Scott's dismounted horsemen. Birney sent fifty mounted troopers from the Seventy-fifth Ohio in a saber charge, which they did "in gallant style." The Confederates scattered, some taking refuge in a nearby swamp. At the North Prong of Black Creek, the Yankees constructed another bridge, and soon began moving toward Baldwin.[82]

McCormick had determined to defend Baldwin; he assembled his troops and received several artillery pieces to strengthen his position. As Birney's column advanced directly toward the railroad junction, McCormick received intelligence that Union soldiers were behind them and approaching fast. Birney's unit from Nassau County had arrived. McCormick and his men seemingly panicked. Birney reported, "During the night and early dawn the rebels evacuated Baldwin and Camp Milton, passing northwestward over Brandy Creek and the St. Mary's, throwing away property in their rapid flight." (Many bluecoats found it amusing that a great number of New Testaments were found discarded along the Confederates' line of flight.) The Yankees destroyed bridges, rail lines, and two boxcars filled with war materials. Inexplicably, Birney then decided to fall back to Magnolia. A recent historian lauded that decision, but a Union outpost at Baldwin would have allowed the

[81] *ORA*, ser. 1, vol. 35, pt. 1, 419–21; "A Yankee Raid in Florida," *Augusta [GA] Daily Constitutionalist*, 10 August 1864, 4.

[82] *ORA*, ser. 1, vol. 35, pt. 1, 419–23; "Correspondence from Hilton Head," *Butler [PA] American Citizen*, 10 August 1864, 4. The writer of the article, a Union participant, gives a vivid account of Birney's raid on Baldwin.

Federals to block one of the major routes used by Confederate cattle drovers. Birney had, in fact, missed a golden opportunity to cripple the Confederacy's Florida supply line.[83]

Brig. Gen. James K. Jackson, who had recently replaced Patton Anderson as commander of the Department of Florida, flew into a rage when he received news of the evacuation of Baldwin. McCormick and his troops had performed with considerable timidity during Birney's raid, missing a golden opportunity to seriously damage Union operations in Northeast Florida. Jackson replaced McCormack, at least temporarily, with Maj. Scott, of the 5th Florida Cavalry.[84]

Cow Country and South Florida

The situation appeared bleak for Confederate sympathizers in South Florida during the first six months of 1864. The Union military presence in that region grew stronger with each passing day while the government in Richmond seemed intent on exacerbating the already difficult situation. Pleas—demands, actually—arrived weekly for more Florida cattle, but on 17 February 1864, the Richmond politicians voted to end the draft exemption for cattlemen and drovers. That unwise decision prompted many Florida cowmen and their families to leave their homes and take refuge in the Union controlled area south of the Caloosahatchee River.[85]

The Federals had long been aware of the importance of South Florida beef, but on 21 February 1864, the *New York Herald* published a supposedly secret document from Rebel commissary officer Pleasant W. White to various Florida cattlemen and military officers. The memorandum issued an impassioned plea for the state's citizenry to send cattle and foodstuffs to the starving Confederate armies. White wrote, "Florida has done nobly in this contest. Her sons have achieved the highest character for their state, and [have] won imperishable honors for themselves. These brave men are now suffering for want of food. Not only the men from Florida but the whole army of the South are in this condition. Our honor as a people demand that we do our duty to them. They must be fed." Maj. J. F. Cumming, the chief commissary officer for Gen. Braxton Bragg's army, reiterated White's plea, bluntly

---

[83] Ibid.

[84] *ORA*, ser. 1, vol. 35, pt. 1, 436; Schafer, *Thunder on the River*, 227–29.

[85] *ORA*, ser. 1, vol. 35, pt. 1, 436.

stating, "I cannot too strongly urge upon you the necessity, yes, the urgent necessity, of sending forward cattle promptly."[86]

The Yankees moved quickly to shut off the flow of beef from South Florida. Both the Battle of Olustee and Birney's raid on Baldwin represented Union attempts to establish a permanent outpost in the East Florida interior. Union garrisons at Gainesville, Lake City, or Baldwin would allow the Federals to block the primary routes used to deliver cattle to Georgia, South Carolina, and Virginia. A Union army veteran writing to a Vermont newspaper stated, "[I]f the Federal troops could succeed only in cutting off the rebel supply of beef from this state, it would be a serious blow to this rebellion.... But we are sorry to record that thus far success has not crowned our efforts in these parts."[87]

As if the residents of South Florida did not have enough problems, deserter bands continued to grow in size and daring. For example, in February 1864 a turncoat troop ambushed a wagon train on the way to Tampa. Capt. John Rodgers Adams, commanding a supply depot at Gainesville, had sent a wagon train filled with food and supplies, guarded by a militia unit, for destitute Confederate sympathizers in South Florida. As the caravan approached the Hillsborough River, at Flat Ford, a large deserter band struck from concealment, taking the whole train. The South Florida Rebels realized that they had to do something to halt the depredations of the turncoats and outlaws but lacked men and weapons to accomplish that objective. On the other hand, the South Florida Unionists also recognized the potential value of those groups. They had already begun organizing a force of Unionists, deserters, and refugees in an effort to halt the flow of beef to the Southern armies. From the small group of "Florida Rangers," the unit had rapidly grown in size until, in February, it received the official designation of the Second Florida (US) Cavalry.[88]

During the early months of 1864, many South Florida cattlemen had become reluctant to accept Confederate paper money in exchange for their beeves. The scrip had value only if the Confederacy won the war, and victory

---

[86] C. Brown Jr., *Florida's Peace River Frontier*, 156–58; Coles, "Far from Fields of Glory," 256–59.

[87] "The Food Situation in Florida," *New York Herald*, 21 February 1864, 5.

[88] C. Brown Jr., *Florida's Peace River Frontier*, 157; Coles, "Far from Fields of Glory," 256–59; "A Letter from the Department of the South," *Rutland [VT] Weekly Herald*, 8 September 1864, 2.

appeared more and more unlikely. News of the debacle on Missionary Ridge apparently trumped recent Rebel victories at Olustee and the Trans-Mississippi Red River Campaign. Early in 1864, "meeting[s] were being held [by South Florida cattlemen]...aimed at requiring the Confederate government to pay for beef with only silver and gold coins."[89]

Gov. Milton was not insensitive to the plight of his constituents in South Florida. Twice he had sent regular troops to the region, and each time events in North Florida had forced their recall. In the initial expedition, the chief executive had dispatched Lt. Col. T. W. Brevard and his Partisan Rangers, along with Capt. J. J. Dickison's company to cow country; but the with arrival of Union troops prior to the fight at Ocean Pond, Milton directed them to return as quickly as possible. The guerrilla chieftain wrote, "Captain Dickison was...ordered to Fort Meade to act in concert with Col. Brevard, as the enemy was in considerable force in the neighborhood of Fort Myers. Just before reaching Fort Myers, orders reached us at headquarters to return, in anticipation of the battle of Olustee." Hurrying the entire distance—six hundred miles—they missed the fight near Lake City by twelve hours. After the Battle of Olustee, the second effort to establish a strong Confederate presence in cattle country involved sending Brevard's unit and two regiments of Georgia regulars to the area. Before they reached Tampa, Milton received orders to send both of the Georgia Infantry units to Virginia. Milton still had Maj. Augusté Bonaud's Battalion, but "[Bonaud] had been diverted somewhat from his direct course to the southern coast for the purpose...of cutting off or punishing a raiding party from St. Augustine, who had made their way southward on the east side of the Saint Johns...for the purpose of driving off cattle and negroes." By the time Bonaud's unit returned, orders awaited the Georgians to return to their home state.[90]

Pres. Jefferson Davis finally took a couple of measures to improve the dismal situation in Florida's Cattle Country. First, he returned "160 men, mostly from South Florida...from the Army of Tennessee to assist with the [cattle] drives; their commander, at the request of commissary agent McKay, was James McKay, Jr. This move paid dividends for the shipment of beef almost immediately." Even the *Philadelphia Inquirer* took note of the change

[89] Proctor, *FL100*, 3 February 1864, 1; Buker, *Blockaders, Refugees, & Contrabands*, 92, 106–108. See generally, Cash, "Taylor County History and Civil War Deserters," 28–58.

[90] C. Brown Jr., *Florida's Peace River Frontier*, 157.

on Florida's cattle frontier. "[D]uring the last six months, parties of Rebel Cavalry, accompanied by 'cow drivers,' have been gathering up beeves and removing them…to Georgia, at a rate of one thousand a week."[91]

The "parties of Rebel Cavalry, accompanied by 'cow drivers'" may be the first recorded mention of "the Cow Cavalry." Charles J. Munnerlyn commanded the unit, officially designated as the First Florida Special Cavalry Battalion. Munnerlyn, a Georgian, had served in the Confederate army, and then as a member of the Confederate Congress, before accepting command of the new unit. Maj. William Footman, a veteran of service in the 2nd Florida Cavalry (CSA), served as Munnerlyn's second in command. Munnerlyn wisely divided his battalion into companies who exercised control over specified counties:

> "Company commanders [and] their area of operations included John T. Leslie, Hillsborough County; Frances A. Hendry, Hillsborough, Polk, and Manatee counties; Leroy G. Leslie, Hillsborough and Hernando counties; William B. Watson, Orange and Volusia counties; Samuel Agnew, Hernando, Levy, Marion, and Sumter counties; James Faulkner, Taylor and LaFayette counties; J. C. Wilcox, Madison County; and Edward J. Ludderloh, Levy and Alachua counties."

A couple of the units, like Ludderloh's, served primarily as home guards, but the new battalion's primary purpose involved protecting the cowmen supplying beef to the Confederate army. Guarding the region from deserters and raiders came a distant second to their principal objective.[92]

The task facing Munnerlyn might have proven too daunting for anyone less committed to the cause. According to Theodore Leslie, "The people [in cow country]…were about equally divided in [Yankee and Rebel] sentiment and action…. The few soldiers stationed here [South Florida] had been indifferently armed…and the supply of ammunition scant." Making matters worse, the politicians in Richmond understood nothing about driving seasons, the difficulties of herding half-wild steers, or the deeply divided nature of the South Florida populace. They wanted a steady flow of beef to the army

---

[91] Ibid.; M. E. Dickison, *Dickison and His Men*, 48; *ORA*, ser. 1, vol. 35, pt. 1, 373; Hewitt, "Forgotten Captive," 46–47.

[92] C. Brown Jr., *Florida's Peace River Frontier*, 157; "Supplies of Beef Furnished to the Rebels from Florida," *Philadelphia Inquirer*, 22 March 1864, 3.

and cared little for the problems faced by the Cow Cavalry and their families.[93]

The Unionist Floridians had initially taken refuge on Useppa Island in the mouth of Charlotte Harbor. They used Useppa as a staging area for operations on the mainland, with the East Gulf Blockading Squadron providing protection for their Union allies. Federal Brig. Gen. Daniel P. Woodbury soon realized that the Florida bluecoats needed a permanent base inland and sent a small unit of Tory cavalry to occupy Fort Myers. When the Union horsemen reached the abandoned Seminole War fortress, they surprised three Rebels trying to burn down the bastion. The arsonists fled as the much larger force arrived. Rodney E. Dillon Jr. reported, "The fort consisted of numerous buildings, including a hospital, commissary building, barracks, bakehouse, wharf, and two guardhouses." In short order, the bluecoats further strengthened Fort Myers by constructing an outer wall that measured "seven feet high and approximately fifteen feet wide at the base." The initial military contingent at Fort Myers included the tiny group of Florida Rangers, and troops from the Forty-seventh Pennsylvania Volunteers, directed by Lt. James F. Myers. Capt. Henry A Crane, commanded the Florida Yankees.[94]

The initial engagement (other than the Christmas foray along the Myakka River) began along the banks at the Caloosahatchee River in early January. "The first recorded combat of the south Florida 'Cattle Wars' occurred at...Fort Thompson, an old Seminole War outpost...where a Union scouting party from Fort Myers tangled with a band of Confederates." Both sides fired a few shots and retreated, with no casualties reported for either side. In some ways, this typified the "combat" in the remote region. A historian summed up the war in the state's lower peninsula, writing, "There were no battles and few heavy skirmishes." Despite the lack of bitter warfare, the conflict brought a great deal of destruction and misery for the area's citizens. The

[93] Coles, "Cattle Wars," 260–62; R. Taylor, "Cow Cavalry," 201–10; Bamford and VanLandingham, "Now Just Hold Your Cow Cavalry Horses"; "Judah P. Benjamin as Cook," *Charlotte Observer*, 10 June 1917, 1. This article is basically a short biography of Munnerlyn, from information supplied by Munnerlyn's daughter.

[94] Coles, "Cattle Wars," 261, quoting from Lesley, "The Organization of the Confederate Cattle Battalion of Florida"; Stone, "Civil War in the Lower Peace River Valley."

bluecoats commenced a systematic strategy "to steal or slaughter all the cattle" they could find.[95]

On 31 January, Capt. Richard A. Graffe sent a small squad of his Pennsylvanians to nearby Fort Denaud. Three days later a force of about forty Rebel horsemen opened fire on four or five men at the stronghold. A company of the Forty-seventh Pennsylvania heard the gunfire and rushed to the aid of their comrades. Seeing the arrival of the reinforcements, the graycoated horsemen retreated. A New York newspaper wrote,

> Capt. Crane and Lieutenant Dening started from Fort Myers...to [Fort] Denaud, on the mainland, with forty men, where our force had a picket of four men, who were attacked by a force of forty-five cowboys, mounted as irregular cavalry. One of our pickets fired, to give the alarm, when the fire was returned, slightly wounding one of our men, a refugee named Whitten. The pickets were driven in a short distance, when the forty men from the fort came up and drove back the rebels.

The bluecoats followed the retreating Confederates on foot, through the darkness. When the Yankees drew up in line of battle the next morning, the Cow Cavalry troopers galloped away.[96]

James D. Green, a popular and respected citizen of Polk County, took refuge at the Union encampment at Fort Myers on 10 March; and Crane immediately assigned him to take charge of the Florida Rangers. Ten days later Green's unit made a foray into the area around Fort Meade. This time, it was no impromptu affair. They went first to the ranch of Willoughby Tillis, an early settler of South Florida who, a few years previously, had survived an attack by Billy Bowleg's Seminole braves. Luckily, Tillis was not home, but the native bluecoats "took all his horses and wagons—his Negro men and all firearms." Next they "visited" the residence of Thomas Underhill. Underhill, a dedicated Rebel, strongly objected to the "confiscation" of his property, and Green's raiders shot him dead. The pillagers' haul included Tillis's slaves, twelve fine horses, two mules, and a number of recruits for the USCT.

---

[95] C. Brown Jr., *Florida's Peace River Frontier*, 158–59; Dillon Jr., "The Battle of Fort Myers," 28. The people who erected the Fort Thompson historical marker chose to ignore the earlier skirmish along the Myakka River.

[96] Buker, *Blockaders, Refugees, & Contrabands*, 148; Harris, "When War Came to Polk County," 28–29.

Bolstered by their early success, the Federals at Fort Myers stepped up their incursions into Confederate territory.[97]

About two weeks later, Green led a second raid on Fort Meade. (The raiders included former Fort Meade residents William McCullough and William McClenithan, who had been relentlessly harried by Southern conscription officers, before taking refuge at Fort Myers.) The bluecoats carried with them an inventory of well-known Confederates they were tasked with killing or capturing. The "hit list" included Willoughby Tillis, James Lanier, Francis Hendry, Jacob Summerlin, F. C. M. Bogges, James R. Durrance, Henry Seward, and Streaty Parker.[98]

On the morning of 20 February, a force of about fifty Federals left Fort Myers headed for Fort Meade. A Confederate scout spotted them near Bowlegs Creek. (Numbers in this fight vary from gross exaggeration to pure fantasy. The Confederates claimed they faced one thousand enemy soldiers; the Unionists estimated they fought at least 150 Confederates. It appears, however, that about seventy Yankees confronted twenty Rebels along the shallow stream.) With McKay and Hendry's men away driving cattle, the Rebel cowmen relied upon a traditional guerrilla tactic: "[B]y skillfully firing and moving short distances, and repeating the maneuver...[the Rebels] even though outnumbered, out-equipped, and out gunned created a deception that hoodwinked the superior force, causing them to retreat." Green and his men quickly returned to Fort Myers, leaving Fort Meade again unscathed. The Unionists could take some comfort in the fact that they dispatched one member of their "kill or capture" list. James Lanier died in the fighting. Henry A. Pine was also wounded in the foot, and the Federals burned the Tillis homestead.[99]

In mid-February, the Forty-seventh Pennsylvania received orders to travel to Virginia, where they would see combat around Petersburg and Richmond. The Second United States Colored Troops replaced the Keystone State soldiers. These African American soldiers had already experienced some of the racism so prevalent at that time—in both the North and South. A historian reported, "On its journey from Washington to New York City...the

---

[97] C. Brown Jr., *Florida's Peace River Frontier*, 159–60; "Important from the Gulf," *New York Herald*, 18 February 1864, 1.

[98] C. Brown Jr., *Florida's Peace River Frontier*, 162–63.

[99] C. Brown Jr., *Florida's Peace River Frontier*, 163; "Historic Peace River Valley, Florida."

Second USCT had to march through Philadelphia to catch a train. An angry mob followed the troops, becoming more unruly as the march continued. One of the soldiers knocked down a heckler for calling his officer a 'white nigger'.... Along the route people expressed their distaste for the contraband regiment by throwing rocks at the train." The situation worsened for the Black troops at Key West, where the regimental chaplain said, "The people [White residents] seem to hate these soldiers simply because they are black." It likely came as a relief to the Black troops when three companies received orders to report to Fort Myers: "From that time on, the Second USCT operated in conjunction with the Second Florida Cavalry [US] during almost all of its combat duty." The Black troops' service, however, would be beset by controversy, which Unionist sources believed resulted from lax leadership and poor military discipline.[100]

On 11 May 1864, the Federals finally captured Tampa. The Yankee force consisted of 140 Florida Unionists, thirty men from the Second USCT, and fifty sailors from the bark *J. L. Davis.* Green took no chances with the operation. He arranged his troops so that the town would be hemmed in on three sides—on the west by the Black troops, who would check any attempt to escape by crossing the Hillsborough River; on the east by the Florida bluecoats; and along the waterfront by the gunboat *Honduras.* The Second Florida Cavalry (US) met a Black man outside of town, who informed them that all Rebel forces had left a couple of days previously for a cattle drive. Despite that intelligence, Green decided to take no chances. They "charged at the double quick, taking the inhabitants by surprise." The first warning many Tampa citizens received came from five-year-old Darwin B. Given, who ran for his home screeching, "The devils are coming! The devils are coming!"[101]

Some of the die-hard Rebels tried to flee Tampa by swimming the river. Two were killed and one wounded in the attempt. Gen. Woodbury reported the Federals took twenty prisoners, but admitted, "only six were soldiers." Tampa had long been a thorn in the Yankees' side, and they took great pleasure in destroying the artillery at Tampa. Fort Brooke consisted of "a single parapet near the water's edge to prevent [an] approach by water," and Green's

---

[100] C. Brown Jr., *Florida's Peace River Frontier*, 162–63; "Historic Peace River Valley, Florida"; Harris, "When War Came to Polk County," 28–29.

[101] Buker, *Blockaders, Refugees, & Contrabands*, 135–38.

men leveled it too. They also captured some antiquated weapons, a sloop loaded with fifty bales of cotton, and ten good horses.[102]

Gen. Woodbury reported that the Black troops "behaved remarkably well," but the "refugee troops, having personal wrongs to redress were not so easily controlled." That turned out to be a remarkable example of understatement. The White invaders went out of their way to hurt and humiliate the citizens of Tampa. They made sure that the USCT soldiers guarded and took the paroles of White residents (which, in a slave-owning society, caused intense mortification). "They [the White Federals] forces on each side of the river...captured the old men and boys in town, all of whom were too old or decrepit, or too young for military service. They ransacked every home, taking and destroying the scant supplies and clothing and everything of value they discovered. Among the places raided was the Masonic Lodge, where they seized the jewels, regalia, and other equipment." (Gen. Woodbury, himself a Mason, discovered those stolen items the following year at Key West and had them returned to the Tampa Lodge.[103])

When the Federals returned to Fort Myers from Tampa, they again received pleas for help from Unionists at Fort Meade. The persecuted loyalists sent word that "enemy troops [Confederates] had been systematically destroying the homes and property of those suspected of Unionist sympathies, as well as rounding up the families of Unionist volunteers and concentrating them in the vicinity of Fort Meade." Additionally, they claimed to have been "cruelly treated." Of course, the bluecoats killed, burned, or stole Rebel property and persecuted pro-Confederate families in their area, but this time it was the Unionists who were suffering. The Yankees decided to halt the mistreatment of loyalist families as quickly as possible.[104]

Two hundred twelve men, about equally divided between the Florida Unionist cavalry and African American troops, rode into Fort Meade on about 24 May 1864. Willoughby Tillis and Lt. Boggess had prepared an ambush, but a captured Rebel scout revealed their plan, and the Federals neatly avoided the enemy's trap: "For the next eleven hours the Union force loaded

---

[102] *ORA*, ser. 1, vol. 35, pt. 1, 389–91; Grismer, *Tampa*, 147–48; "Our Key West Correspondence," *New York Herald*, 20 May 1864, 8; C. Brown Jr., *Florida's Peace River Frontier*, 167.

[103] *ORA*, ser. 1, vol. 35, pt. 1, 389–90.

[104] *ORA*, ser. 1, vol. 35, pt. 1, 389; Grismer, *Tampa*, 147–48; VanLandingham, "The Union Occupation of Tampa, May 6–7, 1864," 13–15.

confiscated forage and provisions and searched the area for Union family members, Confederate sympathizers, and slaves. Their mission accomplished, the Union men put the fort's buildings to the torch." Their haul for the expedition included sixteen new [Black] volunteers, seven Rebel captives, seventy women and children, a thousand head of cattle, and some horses. Although the Rebel cavalry shadowed the raiders at a distance, they apparently did not fire a single shot during the expedition.[105]

The Brooksville Raid

With the successful mission in South Florida, the Union commander at Fort Myers next decided to take the war into the Hernando County interior. On 1 July, a small flotilla of gunboats and transports unloaded 240 troops near Bayport. The strike force consisted of 120 members of the Second Florida Cavalry (US) and an equal number of the Second USCT. Capt. Jonathan W. Childs led the invasion force, with Capt. John Bartholf directing the African American troops.[106]

As it turned out, the raid on Brooksville failed to catch the Rebels by surprise. A few miles inland, the Black troops captured several Confederates without a fight, and shortly thereafter surprised "a Cow Cavalry picket station" twenty miles from the coast. Sgt. Thomas B. Ellis, who had recently returned from service with the Army of Tennessee, decided he could best help his neighbors by delaying the advancing bluecoats. First, he sent a courier to alert the citizens of Brooksville of the impending danger. Next, he assembled a group of local men in a hammock, near a creek, from which they could pour a volley into the ranks of the enemy as they marched along the Brooksville Road. After a brief scouting trip, Ellis returned to find his home guard unit "running all about helter skelter, with no one and everyone in command. Some of them ran back to their plantations to run off their negroes." Recognizing a lost cause when he saw one, Ellis rode west toward Brooksville.[107]

The bluecoats continued along the road to Brooksville sacking homes and burning the abodes of known Rebel leaders. "Some of the Union men, black and white," Filzen wrote, "were natives of the area and familiar with the terrain. They raided their Secessionist neighbors, burning and plundering a swath six miles wide. They confiscated foodstuffs, cotton, livestock, and

---

105 Buker, *Blockaders, Refugees, & Contrabands*, 154–55.

106 Ibid.; Ellis Sr., "Diary, July 1861–April 1865," 8–10.

107 Filzen, "Brooksville Raid"; VanLandingham, "William B. Hooker," 9–10.

took prisoners as they headed north along the path of the Anclote River toward Confederate-held Brooksville." Even the Union officers could not control their troops. Bartholf promised to spare the home of cattle baron and Confederate sympathizer William B. Hooker if his wife would furnish breakfast for the officers. Mrs. Hooker complied with the ultimatum, but Bartholf found his Black troops riding on stolen horses, carrying dresses and female undergarments as they left. Only the intervention of one of Hooker's friends, serving in the Union army, prevented his comrades from burning the plantation house. David Hope, brother of Capt. Sam Hope, and T. G. Leslie were not so fortunate, and their homes vanished in flames.[108]

When they Yankees reached Brooksville they found a ragtag band of militiamen, Cow Cavalry, and local citizens awaiting them on a hill. Long-distance skirmishing ensued, but the bluecoats "seemed more interested in foraging than fighting." Five Rebels and three Yankees died in the fight, and the Yankees began the long retreat to their boats at Bayport. The Hernando County militia timidly followed, skirmishing with the Union rear guard, but the Federal raid produced no permanent damage to the Confederate cause in South Florida.[109]

In the wake of the taking of Tampa, the burning of Fort Meade, and the Brooksville Raid, the Confederates in the Cattle Country began a period of resurgence. Canter Brown Jr. noted, "[The] Confederate cow cavalry companies began to establish their authority, in the Peace River valley and the cattle ranges stretching from the Kissimmee to the Caloosahatchee rivers. By early August fears of imminent attack had diminished to the extent that William J. Watkins, at Bartow, could assert that his family was 'as safe as anywhere in the Confederacy.'"[110]

A large part of the revival of Southern interests stemmed from dissension within the Union ranks. By the middle of 1864, the "honeymoon" between Black and White troops in Federal blue had ended. Capt. Crane wrote,

> It has become really necessary to separate the Col[ore]d Troops from the Refugee families. During our last months absence they [White Union soldiers] have become greatly demoralized and to such an extent has it been carried, that a long continuance can only

[108] Filzen, "Brooksville Raid."

[109] C. Brown Jr., *Florida's Peace River Frontier*, 167–69.

[110] Ibid.

> tend to open irruption & all of this from laxity of discipline that is truly unpardonable. Our [Caucasian] women have been repeatedly insulted—Officers threatened. Horses stabbed with bayonets & otherwise injured. My authority defied by the Guards. My person & house stoned, hissed at, threatened with death & c & this in the immediate presence of an Officer without a remonstrance or attempt to subdue open mutiny to the disgrace of a Military garrison.

Crane requested that the Negro troops be withdrawn from Fort Myers. Campaigning and fighting together should have created an *esprit de corps* within the brothers in blue, but that obviously failed to occur. In an attempt to solve the problem of insubordination, several of the USCT companies were transferred to Cedar Keys, where they would serve with other Tory cavalry companies.[111]

CEDAR KEYS AND LEVY COUNTY

The situation in Cedar Keys and Levy County in early 1864 served as a prime example of the horrors of guerrilla warfare. Neighbors fought neighbors, Whites fought Blacks, and neutrality proved a virtual impossibility. Union gunboats protected the island city, but no place beyond the entrenchments at Station No. Four could be described as "safe." The county contained no large herds of cattle, no real cities except Cedar Keys, with a civilian population of less than two hundred people, and little of strategic importance to the Confederate war effort. A Southern newspaper reported, "the blue Devils have been…making predatory raids [throughout Levy County]," and a local citizen predicted "we [Levy County residents] may be ruined by deserters before the summer ends." The only forces standing between Levy County citizens and the Yankee invaders was Lutterloh's thirty-man Cow Cavalry squad and an old man and young boy militia unit, commanded by Capt. Isaac B. Nichols—the latter company numbering less than fifty "soldiers."[112]

During the early months of 1864, the ships of the Blockading Squadron chased blockade-runners and made occasional forays up the Suwannee and Waccasassa Rivers, but the infantry and cavalry generally avoided the

[111] Coles, "Far from Fields of Glory," 267–68; C. Brown Jr., *Florida's Peace River Frontier*, 171.

[112] "Deserters at Work," *Augusta [GA] Daily Constitutionalist*, 25 June 1864, 4: "Yankees in Florida," *Richmond [VA] Whig*, 24 July 1864, 1.

hinterlands. All that changed in early May with the arrival at Cedar Keys of Maj. Edmund C. Weeks. He brought with him two companies of the Second Florida Cavalry (US) and three companies from the Second USCT. Weeks, born in 1829 at Martha's Vineyard, Massachusetts, had served in the US Navy during the war's first years. By the time he arrived at Cedar Keys to command ground forces, he had apparently become an alcoholic and was driven by an obsession to become a colonel, or better yet, a general. According to one of his officers, Weeks refused to discipline his African American soldiers, even when they took potshots at the fishermen who supplied the Union garrison with seafood. Weeks also halted the search for the Black troopers who tried to sexually assault a White girl.[113]

A naval officer at Cedar Keys noted,

> For the past three days we have received over 300 refugees—men, women, and children. About 100 of these have been formed into a company, and have been fighting the Rebel Cavalry. The [Unionist] people on the mainland are hunted by rebel cavalry and by dogs; their wives are taken prisoner and their houses burned. An old man came off the other day who had his son shot, and we have two young men on board whose father was hung recently, he being a Union man. They put him on a horse, then tying a rope to a tree, fastened to his neck, after which they started the horse. His last words were for the Union.[114]

The Confederates could match, and surpass, any tales of horrors the Federals could tell. Describing a raid by the USCT, a Union officer averred,

> The negro officers and men acted more like savages on this raid than like civilized people fighting for equal rights and national existence. They would rob the poor women of their scanty morsels of provisions and clothes, and if the poor woman should beg for a little to give her children she was sure of abuse.... [When one woman] complained of her treatment, the black devil threatened

[113] Dye, "Defeated in War and Peace," 59–69; see generally, McCullough, "My National Troubles," 59–86. Weeks was a captain when he arrived at Cedar Keys, but received his promotion almost as soon as he arrived in Levy County.

[114] "Florida Refugees," *[Worchester] Massachusetts Spy*, 1 June 1864, 2.

to shoot her with his pistol, and when ready to leave, asked the lady to lay with him.

Other reports told of Unionists killing captives and willful destruction of dwellings and crops. The word "misery," developed a new depth of meaning for the pro-Confederate citizens of Levy County during the last full year of the war.[115]

Maj. Weeks, still seeking glory and military advancement, led a series of raids throughout the region, mainly collecting bales of cotton (which put money in his pocket) and burning plantations. On 6 July, Weeks took two hundred men and moved toward Otter Creek. An informant told him there were "four companies of infantry at Chambers' plantation…and a company of cavalry halfway between the two places." (Almost certainly, Weeks only faced Nichols's old men and boys.) Weeks reported, "I discovered he [the enemy] had a large force, and then fell back to the bayou…. They assaulted us three times with about twice our number, and were handsomely whipped each time." He then sent his Colored Troops to charge the infantry, who reportedly managed to scatter Nichols's old men and boys. The Confederates, who had only three men wounded, fell back to a stronger position, ready to continue the struggle, but Weeks refused to accept the challenge, and retreated to Cedar Keys.[116]

Orange and Volusia Counties

In the 1860s, Orange County, Florida, constituted the outer fringe of civilization on the Central Florida frontier, east of the St. Johns River. It had an 1860 population of only 987 (163 of whom were slaves), but the sparsely settled region was as much a battleground as the rest of the state—and as deeply divided. The Delk family served as an example of the area's deep divisions. William Delk moved to Orange County in the 1850s, with nineteen slaves. Delk freed his slaves just before the war began and became a devout abolitionist, even attending a Loyal Floridian Meeting in St. Augustine in

[115] McCullough, "My National Troubles," 75; "From Florida," *Augusta [GA] Daily Constitutionalist*, 1 July 1864, 2.

[116] *ORA*, ser. 1, vol. 35, pt. 1, 406–408; "Yankees in Florida."

1864. Delk's son, W. P., served in the 8th Florida Infantry, remaining true to the Confederate cause to the very end of the war.[117]

Dickison's capture of the two Union outposts at Welaka and Fort Butler in May 1864, had repercussions for the citizens of Orange County. The Federal troops from St. Augustine returned to the Ancient City, but they intended to exact vengeance on Rebel sympathizers in that area, using fire, death, and terror.[118]

Rather than sending the infantry trudging through a sparsely settled wilderness to be sniped at by Rebel bushwhackers, the Federals decided to use local turncoats and loyalists to do the job. Thomas Petersen led a group of deserters, reportedly numbering seventy-five men, and the Union commanders at St. Augustine ordered Petersen and his group to burn the homes of and kill Clark Stephens, Cornelius Barber, John B. Sykes, James Bradford, and other Confederate sympathizers. Petersen's men did succeed in setting fire to several dwellings, including Winston and Olivia Stephens's Rose Cottage in downtown Welaka. (Clark Stephens, a disabled Confederate soldier, and his wife, Augustina, learned they were on Petersen's "hit list," and fled across the St. Johns at night, finally settling near Ocala.)[119]

Rebel authorities responded quickly to help their Orange County allies. They sent Lt. Melton Haynes, with part of Capt. James W. Starke's Co. H, of the 5th Florida Cavalry, together with a section of Capt. W. B. Watson's Cow Cavalry, to halt Petersen's depredations. A Georgia newspaper reported, "Lt. Haynes, of Capt. Starke's company, with a detachment from Starke's and Watson's cavalry companies, captured eight deserters in Orange County. The notorious [Thomas] Perterman [Petersen] was captured with a muster roll, containing seventy-five names." With their leader in captivity, and their identities known to Rebel authorities, it appears that the deserter band took refuge in Union controlled territory, bringing a measure of peace to Orange County.[120]

Meanwhile, Lt. W. J. McEaddy, J. J. Dickison's second-in-command, took some of Dickison's company, and "paid their respects to the deserters in

---

[117] Porter and Fyotek, *Story of Orlando and Orange County, Florida*, 12; "From Florida," *New York Tribune*, 3 June 1864, 8; Hartman and Coles, *BRF*, 2:836.

[118] Blakely, et al., *Rose Cottage Chronicles*, 341–43.

[119] Ibid.

[120] Ibid.; "More Deserters Captured," *Augusta [GA] Daily Constitutionalist*, 1 July 1864, 2.

Volusia County." Reports had previously reached Tallahassee that "the Yankees have [control over] Volusia County." McEaddy, apparently acting on information from Confederate loyalists, ridded the county of two of the worst Unionist offenders. A pro-Confederate newspaper reported,

> [A] detachment from Captain Dickison's company crossed the St. Johns...and killed two deserters—one by the name of Richardson, who had deserted from Capt. Dickison's company, and the other by the name of Ichabod Turner, who deserted from Capt. Starke's company. Turner was a bad man, and very valuable to the Yankees...Six other deserters were [also] captured.[121]

---

[121] "Captain Dickison among the Deserters," *Augusta [GA] Daily Constitutionalist,* 1 July 1864, 2.

Chapter 7

"Free Men Can Never Be Conquered"

(August–December 1864)

The Situation in Florida

The outlook for Florida's Confederate sympathizers improved only slightly during fall and winter 1864. The transfer of virtually all of the state's infantry companies to Lee's Army had further weakened the tenuous hold the state's Rebels had maintained during the first three years of the war. The Yankees, of course, knew that the Confederacy could not keep their armies in the field without a steady flow of Florida beef, and the Federals redoubled their efforts to halt the shipment of cattle. In mid-August, the Federals made another attempt to establish a permanent outpost in Central Florida. From that garrison the bluecoats could send out patrols to block the various grazing routes used by Southern cowmen. With Florida's increased importance, almost every region of the peninsula—no matter how remote—experienced some type of military action during the closing months of 1864.[1]

Two letters, written on the same day, give a fair indication of the conditions of the contending forces in Florida. Brig. Gen. John P. Hatch, a New Yorker, military academy graduate, Mexican War veteran, who had won the Medal of Honor at the Battle of South Mountain, assumed command of Northeast Florida in July 1864. In August 1864, he wrote to his superior officer, Maj. Gen. J. G. Foster, regarding his plan to bring the entire region from the Suwannee River to the St. Johns under Union control. Predictably, he first requested "three or four more regiments and another battery." Hatch then noted,

> The people [Floridians] confidently expect [Rebel] reinforcements will [soon] be sent to Florida.... The present [Confederate] force is variously estimated at between 1,000 and 2,000 men. The infantry are mostly "new issue" [meaning old men and young teens]. I shall advance the force at Baldwin to Barber's. If you think I had

---

[1] Coles, "Far from Fields of Glory," 256–58; Wynn and Taylor, *Florida in the Civil War*, 84–86.

> better try the raid through Alachua and Marion [Counties], I will by that time be reinforced and will try the thing."

Hatch, if successful, might end the war before the New Year dawned.[2]

On the same day, Brig. Gen. J. K. Jackson, who had recently replaced Patton Anderson as the Rebel commander in East Florida, begged Richmond for additional troops to defend the state. The reply he received stated, "The Major General [Samuel Jones] commanding is not less concerned than yourself in regards to the conditions of affairs in Florida. He fully sees your necessities, but unfortunately has no means of supplying them.... Use the force you have to the best advantage to defeat them [the Federals]. This is all we can do; all that can be expected of you."[3]

GAINESVILLE AND NORTH FLORIDA

Gen. Jackson reasoned that the Yankees' next expedition would be aimed at one of the towns along the Florida Railroad. Anticipating the Union raid, Jackson ordered Capt. J. J. Dickison to leave Palatka and move toward Lake City. The guerrilla leader had not gone far when a courier caught up with Dickison and his cavalry unit, informing them that a large Federal force had again occupied Palatka. Jackson "promptly returned to his Palatka encampment, verified the report, and hurried a dispatch to headquarters, requesting the commanding officer to return his command at once, that he might hold the enemy in check and prevent an invasion of the interior."[4]

While he waited for the return of Co. H, Dickison, with only eighteen men, immediately began skirmishing with the bluecoats. The Rebels assaulted an "enemy force of about 280 men, armed with Spencer [repeating] rifles, six shooting navy pistols and sabers." The Yankees easily drove back the little band of attackers, capturing three men, and scattering the rest. The guerrilla leader, after scouting the enemy, estimated the Unionists numbered "not less than 3,000 or 4,000 men" with artillery. The rest of Dickison's company returned on the evening of 1 August, reinforced by a small

---

[2] Proctor, *FL100*, 4 August 1864, 1; Coles, "Far from Fields of Glory," 227–35.

[3] Proctor, *FL100*, 4 August 1864, 1.

[4] Koblas, *Swamp Fox*, 100–102; M. E. Dickison, *Dickison and His Men*, 76–79.

contingent from Capt. Samuel Rou's Marion County Company and a larger number from Capt. W. E. Chambers's unit.[5]

Before dawn, on the morning of 2 August, a party of Federals, estimated at 280 strong, moved into the countryside, likely to gather supplies, intelligence, and contrabands. Dickison and his men, though greatly outnumbered, pursued the raiders and attacked them, driving the Unionist back toward Palatka and the protection of their gunboats. Dickison wrote, "Charge after charge was made by our brave boys, the enemy fighting and giving way very sullenly. They were six miles from Palatka.... The fight was very hot...[and] at every charge of our dauntless men, they [the Federals] would give way, but soon rally."[6]

As the opposing forces reached "the hill overlooking the city of Palatka," the Union gunfire slackened and came to a brief halt. Dickison, believing that the enemy was preparing to surrender, dashed forward, shouting for his men to cease firing. "Just at this critical moment," Dickison reported, "the enemy opened a deadly fire from their six-shooters, and...Charlie Dickison, the [only] son of the captain fell mortally wounded." "Joe, I am killed," young Dickison called to his friend, Sgt. J. C. Crews. Capt. Dickison hurried to his son's side, took him in his arms, and held his child as the young man died. The Union raiding party apparently used the lull in the fighting to fall back to their lines, and the next day they boarded ships and returned to Jacksonville. The Rebels claimed they killed fourteen, wounded thirty, and captured twenty-eight. Their own losses were reported as one killed and one wounded.[7]

The guerrilla leader carried his son's body to their camp on his own horse, and eventually had him taken home. The ladies of northern Marion County lovingly interred Charlie Dickison at the graveyard of the beautiful Orange Springs Church. It seems safe to say that J. J. Dickison never quite recovered from the loss of his son. He continued his irregular campaign, and became Florida's premier war hero, but in dark hours he likely blamed himself for his decision to call for a cease-fire.[8]

While the bluecoats at Palatka kept Dickison occupied, a large force left Jacksonville, heading up the St. Johns River to Magnolia, in Clay County. It was one part of Gen. Hatch's three-prong plan to establish a garrison in

[5] Ibid.

[6] Ibid.

[7] Ibid.

[8] Ibid.

Alachua County, thereby impeding the flow of cattle to the Rebel armies to the north. Hatch's battle plan envisioned simultaneously sending three columns into North Central Florida, thereby forcing the Rebels to divide their already limited manpower to deal with each battle group. Col. William H. Noble, with two regiments of USCT, a small group of mounted infantry, and three cannons would threaten Lake City. Maj. Edmund Weeks, with a force described by Southern sources as "500 Union men, deserters, and Negroes," would simultaneously advance from Cedar Keys toward Gainesville. While the graycoats shifted troops to deal with Noble and Weeks's columns, Col. A. L. Harris's contingent—comprising 250 men from the Seventy-fifth Ohio Mounted Infantry, Fourth Massachusetts Cavalry, and one cannon from the Thirty-first Rhode Island Artillery—would move overland from Magnolia to the Florida Railroad line, and from there to Gainesville. If all went as planned, this complex campaign would provide the Federals the toehold they desired in Alachua County.[9]

After unloading his men and equipment at Magnolia, Col. Harris followed basically the same route used by Gen. Birney a several weeks earlier. They crossed Black Creek and then marched to Trail Ridge, and from there, to Baldwin. Harris's bluecoats then proceeded to Starke, in Bradford County, where they found several boxcars loaded with commissary stores. The Union soldiers took what they could carry and burned the rest. The Federals then moved south, along the Florida Railroad tracks, bivouacking for the night ten miles from Gainesville. Along the way the raiders burned the plantations of Dr. McRae and Mr. Boulware, as well Boulware's mill and gin house. There, Harris's raiders received reinforcements in the form of a "detachment of Unionist Floridians, totaling altogether 104 officers and men."[10]

News of the movement into Alachua County soon reached Dickison. With a heavy heart he left the body of his son, Charlie, in the care of the ladies of Orange Springs to complete the burial. The guerrilla leader had only 135 men as he followed the trail of Col. Harris's marauders. Near Waldo, however, Rebel reinforcements arrived. Gen. W. A. Owens reached Dickison first with about twenty volunteers from Marion County. Those volunteers,

---

[9] Schafer, *Thunder on the River*, 231–32; "The Latest Disaster in Florida," *[Washington, DC] Daily National Intelligencer*, 7 September 1864, 2; *ORA*, ser. 1, vol. 35, pt. 1, 429–35; Barker-Benfield, "Union regiment's raid from Baldwin proved less than successful."

[10] Ibid.

styling themselves as "Marion Boy Braves," ranged from thirteen to fifteen years of age. Next came a contingent composed of "new issue" militia, but these "cradle to the grave" soldiers had received no training and had few arms. Lt. Melton Haynes came next with about forty veterans from Capt. L. G. McElvey's Co. H, 5th Florida Cavalry Battalion. The 5th Florida horse soldiers had fought with Dickison before, and he knew they could be counted on in the coming battle. All total, Dickison had 290 men (and boys), but only 175 would take part in the actual fighting.[11]

As they had rode south toward Gainesville, the grieving father discussed the situation with Gen. Owens. The older man, who had earlier resigned his military commission due to age and infirmity, later recalled the conversation. Though devastated by the death of his only son, the guerrilla leader still held out hope for the Confederacy. "We will meet the enemy very soon," he said. "We must win this fight or the country [Florida] is gone. I can see in my brave men a determination to sacrifice their lives, or win the fight, and I know they will win it. They have seen their homes invaded, and the sore distress of their hopeless families and neighbors. Such men may be killed, but never conquered."[12]

The Federals reached Gainesville at about 6 a.m. on 17 August. A small party of Confederates fled precipitately at the first shot. This timorous action may have paved the way for Dickison's victory. Col. Harris seemingly decided that the graycoats he had just routed were the only enemy troops in the area, and some of his men began looting nearby buildings. Meanwhile, the veteran commander set out a picket line on the edge of town, emplaced his cannon along the railroad tracks in downtown Gainesville, and established his command post at the rail depot. Gen. Hatch would later lament that Harris lost control of his men and "allowed his men to scatter through the town to 'pillage.'"[13]

The Yankees had occupied Gainesville for less than an hour when two shells from Lt. Bruton's guns announced the arrival of Dickison's men. The Union commander quickly recognized the gravity of the situation and

---

[11] M. E. Dickison, *Dickison and His Men*, 87–96; "History of Gainesville, Florida"; "About General Dickison," Ocala [FL] *Banner*, 30 July 1909, 11.

[12] M. E. Dickison, *Dickison and His Men*, 91.

[13] *ORA*, ser. 1, vol. 35, pt. 1, 429–35; Barker-Benfield, Aug. 9, 1864: Union regiments' raid from Baldwin proved less than successful"; "History of Gainesville, Florida."

hurried to get his men in position to meet the coming assault. The companies of Seventy-fifth Ohio Mounted Infantry were strung out to the left and right of the depot, behind the railroad embankment, and a company of the Fourth Massachusetts Cavalry would serve as a mobile reserve. Harris noted that his right and left flanks were anchored "on a swamp and [a] thicket."[14]

With bugles blowing and screeching a "Rebel Yell," or "Remember Charlie Dickison," Capt. Dickison led his Rebel horsemen in a wild frontal assault. They were greeted by a heavy volley from Harris's bluecoats, but rather than retreating and reforming for a second attack, the Confederate horsemen swung through the openings in the Federals' line, to the right and left, and began encircling their opponents. Harris reported, "The enemy was checked in front, but immediately surrounded me [the Union position] with his whole force, compelling me to send Co. B, Fourth Massachusetts Cavalry, to the rear of the town, and [then] throw portions of the Seventy-Fifth Ohio Volunteer Mounted Infantry, on both the right and left flank, thus weakening my first line." The bluecoats held on until about 9 a. m., when Harris, pressed on all sides, decided to fall back and establish a new line. While justified by the situation he faced, it proved to be a fatal choice. Dickison's men charged into the midst of the retreating Union soldiers, firing with rifles, shotguns, and six-shooters. The Federal troops never had a chance to reform. With graycoats pressing into them from all sides, and no place to take refuge, the Yankees' battle line quickly disintegrated into a disorderly mob.[15]

A Union soldier reported,

> The rebels...circled to the north of town. We were surrounded. Some of our men [were] killed or disabled. Ammunition was fast dwindling. [Col.] Harris called in our line. A complete rout followed. Mounting our horses we backed off into town, then out on the road to the west where the [Fourth] Massachusetts cavalry held the enemy in check for a short time. A woman standing in the doorway of a shack fired at Colonel Harris. In a flash she was shot down. Harris escaped out by the road to the west.

---

[14] "The Latest Disaster from Florida"; "History of Gainesville, Florida"; M. E. Dickison, *Dickison and His Men*, 87–96.

[15] Cox, "Battle of Gainesville; Gainesville, Florida; Florida's Swamp Fox does battle"; M. E. Dickison, *Dickison and His Men*, 87–91.

Another bluecoat wrote, "Colonel Harris, who was near us at the time, said: 'Boys, I am sorry for you; I have stayed by you till the last minute; good bye;' and away he went through the dust on his splendid horse."[16]

The defeat of the larger Union force had been accomplished, but Dickison pushed his men hard to capture the beaten foe. A veteran of the engagement recalled,

> At this time Captain Dickison dashed through the streets, calling for his men to mount their horses and follow, which was quickly done—the enemy scattering along the roads and through the woods, pursued on every side by the brave southern boys. The pursuit continued as far as Newnansville, 15 miles distant, many [Federals] being killed or captured on the road.

Dickison placed the Union losses at twenty-eight killed, five wounded, and 188 captured. Col. Harris, after a difficult journey, eventually staggered into Magnolia a couple of days later. In all, about forty bluecoats made it back to Union lines. The Confederates reported their casualties as two killed and two wounded.[17]

The list of captures might have been greater, but Gen. Jackson, the Confederate commander in East Florida, recalled Dickison before he could finish his pursuit to prepare for the expected column from Cedar Keys. For several days, newspapers and Federal officers had been reporting that a second column, numbering five hundred men, would soon reach Gainesville from Gulf coast city. As a result, Jackson directed Dickison to remain in the area "for a few days to watch for a demonstration from Cedar Keys."[18]

Somehow the ballyhooed invasion from Levy County never materialized. Records about the operation are almost nonexistent, except for a couple of paragraphs in obscure journals. A Georgia paper reported, "Judge Dawkins, at Gainesville, says the negroes and deserters from Cedar Key have been driven back by Capt. Lillenthal [Ludderloh], who killed one and captured another. They have gone back to Cedar Key." If the infliction of two

---

[16] Koblas, *Swamp Fox*, 107; Coles, "Far from Fields of Glory," 229; William B. Southerton reminisces, Ohio Historical Society, Columbus, OH; "Prison Reminiscences of George H. Luther," in *Shot and Shell*, 271.

[17] "Battle of Gainesville; Gainesville Florida"; "Florida's Swamp Fox does battle"; Koblas, *Swamp Fox*, 106.

[18] *ORA*, ser. 1, vol. 35, pt. 1, 436; Proctor, *FL100*, 22 August 1864.

casualties by an enemy force of perhaps thirty-five men caused five hundred Federals to retreat to the safety of their gunboats on the coast, it must rank as one of the most amazing feats in the annals of warfare.[19]

Gen. Hatch's well-conceived plan had accomplished very little to aid the Union cause, and the commander in Northeast Florida immediately sought scapegoats for the failure. He blamed Col. Harris, claiming he allowed his men to scatter to the town to "pillage." There was some looting, but Harris seems to have had virtually all his soldiers in line when Dickison attacked. The defeat apparently occurred because the Rebels did not fall back and re-form for a second frontal assault, as Harris expected. Instead, they galloped through the enemy defenses, and Yankees were soon taking fire from the front and rear. That caused the bluecoats to panic.

Hatch, however, saved his bitterest invective for Col. Noble. In hindsight it seems clear that Hatch intended for Noble to attack (or at least threaten) Lake City. Instead, the Ohio leader conducted a raid through the Bradford and Columbia Counties, burning cotton, plantations, gin houses, and taking enough African American males to form a company for the USCT. Hatch's orders, however, had not specified that he expected an attack on the Columbia County seat rather than a foray into the countryside, so the general could not, in fairness, blame his subordinate for failing to read his mind. If Hatch commented on Maj. Weeks's abortive expedition from Cedar Keys, it is not included in the *Official Records*.[20]

CLAY COUNTY

Gen. Hatch next decided to beat the guerrillas at their own game. He wrote, "I am now building a post at Magnolia, [with] a small but strong fort. It was my intention to put Colonel [James] Montgomery's regiment there, and have him bushwhack. He understands the business, and assisted by a company of Floridians, would do more [than anyone else] to keep the enemy from taking the offensive." Citing ill health, however, Montgomery retired from the service and returned to Kansas. In his absence, the war continued along the St.

---

[19] *ORA*, ser. 1, vol. 35, pt. 1, 37 and 436; Proctor, *FL100*, 26 August 1864; "Gov. Milton," *Augusta [GA] Daily Constitutionalist*, 12 August 1864, 1.

[20] *ORA*, ser. 1, vol. 35, pt. 1, 427–35; 303–304; Schafer, *Thunder on the River*, 232.

Johns, with numerous skirmishes occurring in the Clay County countryside.[21]

Dickison and his men got most of the credit for the fighting around Magnolia, but there were several units that kept the bluecoats largely confined in their enclave along the St. Johns River. Col. Noble provided a derisive (but basically accurate) description of the Rebel guerrilla warriors he faced. "The cavalry," he penned,

> were a body of mounted Floridians, commonly called 'crackers;' and Falstaff's men in buckram could form no comparison to them in appearance. They were all sorts and sizes, and arrayed in a kind of homespun disuniform, from gray-black to butternut, and all intervening shades, mounted on horses ranging from the pony, weighing about as much as his rider, up to a sizeable animal. Modern times have seen few such cavalcades.

Despite their variegated appearance, the Rebel guerrillas continued to hold the bluecoats at bay in Northeast Florida.[22]

When the raid from Cedar Keys failed to materialize, Dickison and his men moved to Timmons Crossroads, in the Etoniah Scrub. They briefly basked in the glory of their victory at Gainesville, which did much to raise the spirits of Florida's Rebel sympathizers. A local newspaper gushed, "The result of the Battle of Gainesville is cheering to the hearts of people.... It is striking, a noble illustration of the fact that a people so zealous in the defense of their rights, so resolutely determined to live as free men, can never be conquered."[23]

On 23 October, Dickison received a request for aid from Lt. Melton Haynes, of the 5th Florida Cavalry. Haynes, with perhaps 150 men, had been attempting to keep three to four times that number of bluecoats confined at their Magnolia base. The Federals first moved north to destroy a Confederate battery on Fleming Island established five months previously as a way to "harass Union ships" on the St. Johns River. With that accomplished, the bluecoats began moving into the interior to capture—or destroy—Rebel supplies and recruit African Americans for their army. An early historian of the state's

---

[21] Proctor, *FL100*, 22 August 1864; Phillips, "Montgomery, James."

[22] Croffut and Morris, *Military and Civil History of Connecticut*, 733; "Calendar of Civil War Activity 1862–1864"; *ORA*, ser. 1, vol. 44, 824–25.

[23] M. E. Dickison, *Dickison and His Men*, 104–106.

conflict accurately described the region's situation as "a dismal series of forays, night marches, surprises, captures, skirmishes, burning, pillagings, robberies, murders, strokes and counterstrokes in active guerrilla warfare, with now and then a skirmish that partook of the character of a formal battle."[24]

Dickison wasted no time in accepting Haynes's offer and immediately began making life miserable for the enemy in Clay County. The partisan command, including Haynes's troopers and a detachment from the company formerly directed by Capt. W. E. Chambers, likely totaled two hundred men.[25]

The "Swamp Fox" of Florida soon received intelligence that a Federal raiding party was returning from a foray on Middleburg. About 55 men of the Fourth Massachusetts Cavalry galloped back after sacking and burning the village. A historian recorded, "The raiders [had] burned warehouses and other buildings, and looted the remaining houses and businesses." Dickison intercepted the returning pillagers near the Halsey Plantation, at a place called variously "Big Gum Creek" or "Big Gum Swamp." Facing the Rebels, the Bay State troopers "charged with drawn sabers; the polished blades flashing in the sunlight." A volley from the Confederates emptied a couple saddles and disrupted their battle line. Dickison, with some exaggeration, wrote, "As they [the Federals] drew near, they were met with a telling volley. They halted, but quickly formed their line and charged again; our men meeting them with a heavy fire. Our artillery then opened upon them, and they fell back in great confusion, our intrepid men charging them, killing and capturing the entire command as they retreated." The graycoats claimed they killed, wounded, or captured thirty-five bluecoats; while the Federals admitted a loss of twenty-nine men. Dickison's loss were a few horses killed or disabled.[26]

Not all the fighting at this stage of the war involved artillery or flashing swords, but it sometimes devolved into bushwhacking and indiscriminate slaughter. For example, a small group of Clay County teenagers, led by George Washington Bardin and Moses Joiner, slipped up on a sentry post,

---

[24] Blakey and Deaton, *Parade of Memories*, 83; M. E. Dickison, *Dickison and His Men*, 107; Schafer, *Thunder on the River*, 230–31; W. W. Davis, *Civil War and Reconstruction in Florida*, 297–98.

[25] M. E. Dickison, *Dickison and His Men*, 104–107; Blakey and Deaton, *Parade of Memories*, 83.

[26] M. E. Dickison, *Dickison and His Men*, 107–109; Blakey and Deaton, *Parade of Memories*, 87.

manned by Black troops, and murdered several. It did not take long for the Unionists to discover the identity of the culprits, and they sought revenge on the teens. Luckily, Bardin's father was a member of a state militia company, and he quickly sent the youngsters to Hamilton County with a request that the captain keep the youths out of harm's way until the war ended. Actions such as Bardin's and Joiner's unprovoked attack incensed the Yankees, Black and White, and many concluded that killing, without quarter, best suited such vile opponents.[27]

Col. William H. Noble, who had often been a thorn in the side of Florida's Confederates, ended his military career in a most ignominious fashion. On 22 December 1864, he and two companions were returning to St. Augustine after attending a court martial in Jacksonville. He refused a guard, apparently deciding that the route east of the St. Johns had been cleared of enemy combatants. A period historian recorded, "When he had ridden about half the distance, three rebels dashed out of the woods in front and rear and made him a prisoner." The graycoats also captured Capt. Young, of Gen. Birney's staff, and Lt. Rice, Florida provost marshal. Noble eventually was sent to Andersonville, the hell hole of the Rebel prison system.[28]

William G. Ponce, a member of Co. A, Second Florida Battalion, had led the tiny operation that captured Noble and his companions. Wounded at the Battle of Cold Harbor, in Virginia, Ponce had joined a river watch unit at Doctor's Inlet while he recuperated. Somehow Ponce had gotten wind of Noble's route and convinced a couple of friends to join his operation. He handed over his prisoners to a Confederate officer and returned to his comrades at Doctor's Inlet. The bluecoats would raid the watch unit in January 1865, specifically to take Ponce prisoner. Noble survived his incarceration at Andersonville, but Ponce died in the Union prison at Hilton Head, South Carolina.[29]

ST. AUGUSTINE

Recent historians have described St. Augustine, during the war's last months, as "a rest camp for the Department of the South" and "peaceful." The bluecoats had finally achieved a modicum of control over the region, but just as

---

[27] Schafer, *Thunder on the River*, 230; Rerick, *Memoirs of Florida*, 1:428–29. Postwar, Bardin served for many years as a Clay County judge.

[28] Hartman and Coles, *BRF*, 3:1175; "The Civil War in Clay County."

[29] Ibid.

it had been during the previous two years, the outskirts of the Ancient City could also be a very dangerous place. In early November, J. J. Dickison received intelligence from his scouts in St. Johns County that large companies of Federal soldiers had begun making daily visits to the Fairbanks house, only two and a half miles from the city walls. (No reason has been discovered for the continued popularity of the Fairbankses' dwelling—whether it was women, liquor, or nobler pursuits is unknown.) Dickison determined to make the bluecoats pay for their lack of caution. He took fifty men from Chambers's company, and the same number from his Co. H, and in the inky darkness crossed the St. Johns. The Rebels had only a single flatboat, so it took several trips to transport one hundred men and horses across the waterway.[30]

Capt. Dickison arranged his troops carefully. He divided "his command, leaving a detachment in front of St. Augustine to guard against the enemy coming out [once he sprang his trap] from that point." Lt. Reddick led that unit. Reddick would allow the Federals to pass through his position for their daily visit to Fairbanks, and then assault them from behind when Dickison attacked those already at the Fairbanks residence. In that way, the "Swamp Fox" would bag the entire column.[31]

Dickison described the action:

> Captain Dickison concealed himself about twenty feet from the road, and allowed the [Federal] advance to come within a few yards of him. He then arose and ordered them to surrender. They replied with a volley into his command, which drew the fire of our men [Confederates] who were dismounted. A charge was immediately made which resulted in killing and capture of the advance. The Federal battalion wheeled around and dashed back in great confusion. Lieutenant Reddick bravely charging them through an impenetrable scrub, killing three and mortally wounding their commanding officer, the major of the battalion.[32]

Dickison asserted that the Rebels captured thirty-five prisoners. He makes no mention of the dead, but it seems they also killed four bluecoats.

---

[30] M. E. Dickison, *Dickison and His Men*, 109–12; Koblas, *Swamp Fox*, 114–15.

[31] Ibid.

[32] Ibid.

The Confederates reported no dead or wounded. Dickison and his men, with their prisoners, re-crossed the St. Johns the following day.[33]

COW COUNTRY AND SOUTH FLORIDA

The bitter schism between the USCT and Florida Unionists proved extremely beneficial to the Rebel cause in South Florida. Nothing but a few minor skirmishes occurred there during the final months of 1864. As a result, the Cow Cavalry continued collecting and driving herds to the towns and railheads in South Georgia. Col. Munnerlyn boasted that the Cattle Battalion "had caused the federal troops to virtually cease cattle confiscation in the area." However, it was the friction between White and Black Federals that helped the Confederacy more than the revolvers of the Southern cowmen.[34]

The Federals did score a minor triumph in South Florida, in September, though they did so as a result of a mistaken identity. A Northern journal reported that Master Henry B. Carter, with about ten men, proceeded up the Manatee River to destroy "a sugar mill owned by [Confederate president] Jeff Davis." Maj. Robert Gamble, who had come to the state to fight in the Second Seminole War, had built a plantation house on the Manatee between 1845 and 1850. The dwelling was constructed of tabby and had walls almost two feet thick to withstand the ravages of hurricanes and Native American attacks. Gamble's slaves had also dug almost sixteen miles of ditches to irrigate his sugar cane fields. He had also located on the plantation grounds a mill to convert his crops into sugar, syrup, and molasses (which could easily be converted into rum). Deep in debt by 1858, Gamble sold his holdings to Robert M. Davis, of Louisiana, but the rumor spread throughout the region that Pres. Jefferson Davis actually owned the property.[35]

The Federals knew that the Davis mill had produced fifteen thousand hogsheads of sugar the previous year, which had been sold to the Confederate Commissary. Master Carter and his men placed artillery shells in the refinery and house and then detonated them. The mill was completely destroyed, but very little harm was done to the Gamble Mansion. Before they left the area, they also demolished a second gristmill on the south side of the river. Grismer

---

[33] Ibid.

[34] C. Brown Jr., *Florida's Peace River Frontier*, 171; Coles, "Far from Fields of Glory," 267–68.

[35] "Washington," *Philadelphia Press*, 12 September 1864, 2; Cox, "Gamble Plantation Historic State Park."

wrote, "The loss of the grist mill proved a particular hardship for area residents because…all the families in Manatee [County] were dependent upon it for [grinding] their grits and cornmeal."[36]

The Cow Cavalry carried out one minor operation during this period that caused some consternation among the enemy but did nothing to change the military situation. On 27 August, a picket stationed along the south bank of the Caloosahatchee River, saw a "signal of distress," from the Confederate side. The Union guard recognized "a negro well known to him was standing awaiting the boat." Ignoring the normal procedures, two guards—a White and Black man—pushed directly across the waterway to rescue the "contraband." According to Capt. Crane, "Instantly a heavy volley of musketry was poured into them, killing both [men]." The Rebels then turned their guns upon some nearby Union sloops. Total casualties in this little ambush were three Federals killed and one wounded.[37]

Crane's closing comments in his report on that incident revealed how deep the divisions had become between the White and Black soldiers. He again begged for the removal of all the Black troops and return of the Second Florida Cavalry [US] companies from Cedar Keys. "I am fully satisfied," he wrote, "that each [race] should be separate to accomplish anything…. Our recruiting [of Whites] has been killed off almost entirely, and desertions have commenced." He had only forty White soldiers under his command, he explained, because the Black soldiers left due to their "impudence."[38]

During late August or early September, an event occurred in Manatee County that remains a mystery. What is known is that…

> in late summer of 1864 a little group of [Confederate] deserters from Fort Brooke, accompanied by some escaping slaves, made their way through the [Gulf Coast] jungle to the mouth of Anclote River, hoping to be picked up by a Federal gunboat. To attract a ship, they built a huge bonfire on the highest point of land, not [realizing] they were [being] pursued.

[36] Cox, "Gamble Plantation Historic State Park"; Grismer, *Tampa*, 148.

[37] Buker, *Blockaders, Refugees, & Contrabands*, 152; *ORA*, ser. 1, vol. 35, pt. 1, 614.

[38] Ibid.

A party of South Floridians was close on their heels. "Waiting for low tide, they [the pursuers] crossed the river at night, surprised the fugitives, and hanged them all."[39]

For almost a century, a search for the identity of the killers has raged. For many years local historians suspected Rebel Capt. Sam Hope, possibly because, postwar, he was one of the first settlers along the Anclote. However, Dr. Joe Knetsch has since proven the impossibility of Hope's participation in the hanging, as he was serving in the Army of Northern Virginia during that entire time. It now seems likely that some unidentified guerrilla band or militia unit killed the deserters and contrabands. (In one of those quirks of history, the site of the lynching became a favorite picnic spot for Pinellas County residents during the early twentieth century.)[40]

CEDAR KEYS AND LEVY COUNTY

The situation in Cedar Keys and Levy County during the last four months of 1864 could hardly have been worse for the Federal cause. Maj. Weeks's timidity in combat, as evidenced by his half-hearted effort to support Col. Harris's Gainesville raid, did little to help the Union cause in the region. In September, Weeks, in a drunken stupor, shot and killed a Yankee sentry and was taken to Key West to stand trial for murder. As a result, Federal leadership at Cedar Keys deteriorated even more. The men of the Second Florida Cavalry (US) elected Capt. John D. Green, a die-hard Florida Unionist, to replace Weeks. When the results of the vote were forwarded to Key West for approval, both Capt. Jonathan W. Childs, commanding at Ft. Myers, and Col. B. R. Townsend, leader of the Second USCT, strenuously objected to Green's appointment. Gen. John Newton apparently never found an appropriate replacement until Weeks returned, and the Union military in Levy County experienced a period of stagnation.[41]

Aside from the raid toward Gainesville, a naval expedition up the Suwannee River to Clay's Landing is the only known military during the last six

---

[39] "Footnote to the Civil War," clipping file at the Tarpon Springs Area Historical Society, Tarpon Springs, FL. Author and date unknown, Zack C. Waters collection, Rome, GA.

[40] Ibid.; Knetsch, "Forging the Florida Frontier," 35.

[41] Buker, *Blockaders, Refugees, & Contrabands*, 163–64; Dye, "Defeated in War and Peace," 49–52; "Capt. James Dopson Green, Company B"; C. Brown Jr., *Florida's Peace River Frontier*, 163–69.

months of 1864. The Rebels had gathered about one hundred bales of cotton there and had constructed an earthwork to protect the valuable cargo. A US gunboat sent into the hinterlands reportedly bombarded the Confederate position, and the graycoats scurried into the nearby woods and swamps. The Rebels fired their precious commodity, but the sailors salvaged what they could of the cotton and quickly departed. The sailors boasted that they killed and wounded several of the enemy, but a report by a Federal officer casts doubt on that claim. Florida Unionist, Lt. William McCullough, who accompanied the expedition to Clay's Landing, noted, "As for the crew, they do not wish to go up the river, and the captain and his officers are cowardly, being strongly tinctured with sissies."[42]

Sadly, those problems paled in comparison to the other difficulties that plagued the citizens and soldiers at Cedar Keys. In November, a Boston newspaper reported, "It is stated that the loyal white refugees at Cedar Key, Florida are suffering terrible privations...and, three hundred and forty out of three thousand have perished from want of the necessaries of life." Sadly, that estimation of the death total appears to be an understatement. Lt. McCullough kept a tally of the dead in his diary (beginning in August), and many days the totals numbered eleven to fifteen deceased. The Union officer's diary entry, dated 3 September, explained the problem: "Still no rations...three days without bread. The troops [are so hungry they] are trying to eat some rotten flour...[T]his flour has been set outdoors on account of the worms and weevils in it. It is the kind that kills people so fast. The flour is bitter as gall. Six of my men are taken sick today from the affects of this. [The men]...have high fevers and vomiting." No accurate numbers of the dead at Cedar Key, during that time, has yet been located.[43]

FLORIDA COMMANDS AND COMMANDERS

In November 1864, the Department of South Carolina, Georgia, and Florida sent Richmond a list of military assets in each state. The Florida tally sheet did little to inspire confidence in the state's ability to hold the vital cattle

[42] "Successful Naval Expedition up the Suwannee," *New York World*, 6 August 1864, 1; McCullough, "My National Troubles," 113. "Sissies," then as now, was a derogatory term for gay men.

[43] "Suffering of Loyal Refugees," *Boston World*, 21 November 1864, 3; McCullough, "My National Troubles," 115–17, provides a fair sampling of McCullough's diary entries on the food shortages.

range in South Florida or its ability to maintain a route to move beeves to the Confederate armies outside the state. So desperate had the young nation's situation become that the Davis Administration could offer only hopes and prayers for Confederate loyalists in the Land of Flowers.[44]

Brig. Gen. William Miller, born in New York but raised in Louisiana, had been a pre-war attorney and sawmill operator in West Florida. He had earned a reputation for bravery at Perryville, Kentucky, and Stone's River, Tennessee, but he suffered a grievous wound in the latter fight that sidelined him for months. He returned to duty in 1864 and had helped to make life miserable for Gen. Alexander Asboth, commander of Union forces in West Florida. Again cut loose from his duties around Pensacola, "[h]e offered his services to Brigadier General John K. Jackson, commanding the District of Florida, with headquarters at Lake City. Jackson, who desperately needed qualified subordinates at the time, accepted him to command Sub-District No. 1."[45]

The 2nd Florida Cavalry (CSA) was Miller's most reliable force. Col. Abner H. McCormick, despite complaints of poor performance by various commanding officers, retained command of the cavalry units. The 2nd Florida Cavalry had ten companies, but half of them served exclusively in the state's Panhandle region. Also available to Gen. Miller were three companies of the 5th Florida Cavalry, directed by Capt. Robert J. Chisholm. Though technically separate entities, the East Florida companies of Chisholm's command had often fought with Second Florida Cavalry to confront a more numerous foe (as they had at St. Augustine and Magnolia). Lt. Gen. William J. Hardee, commanding the East Florida region, did not include the Cow Cavalry on the list of available combat forces, as their primary responsibility involved gathering, and driving, cattle. Fighting came only as a last resort to Munnerlyn's men. They would, as a general rule, skirmish only to protect the herd or their men.[46]

The Florida Reserves, led by Capt. Isaac B. Nichols, constituted Gen. Miller's only infantry unit in East Florida. The Reserves were composed of males, too old or young for service in the Army of Tennessee or Army of Northern Virginia. Despite the poor quality of Nichols's battalions, the commander had a reputation as a hard fighter. The Gainesville *Cotton Plant*

---

[44] Proctor, *FL100*, 20–21 November 1864.

[45] Hewitt, "William Miller," 4:176–77.

[46] Proctor, *FL100*, 20 November 1864; Robertson, *Soldiers of Florida*, 285–96.

opined, "Capt. Nichols, with his 'New Issue,' has captured one deserter in Levy County, and we have no doubt of his capturing many more. There is one thing certain, Old Nick, with his 'New Issue,' will fight, and if he beats some of the 'old issue,' we would not be seriously surprised nor mortally mistaken."[47]

The desperate condition of the Confederate governments in Richmond and Tallahassee showed in a series of actions taken during the last months of 1864. Gov. Milton requested the Davis administration to return Finegan's Brigade from Virginia to Florida if "compatible with the public service." Milton made a strong case for this plea. The Florida troops, he wrote "are weak and debilitated from bad health, from sickness and fatigues of an arduous campaign, and in poor condition to stand the freezing weather of a Virginia winter; and the strong probability of the invasion of the Southern coast of the Confederacy...[and] the Florida troops would be better able to endure the winter in a milder climate, and would greatly recuperate [their] wasted health." With Gen. Robert E. Lee's Army being daily depleted by battle casualties and desertions, he could scarcely afford to lose a company, much less a brigade. The Floridians remained with the Army of Northern Virginia all the way to the surrender at Appomattox Courthouse.[48]

An even greater act of desperation involved the Richmond Government's attempt to enlist African American soldiers for the Rebel army. A memorandum stated, "The Confederate Bureau of Conscription today issued General Circular No. 36, authorizing the impressment of free negroes and slaves into the Confederate army...Florida's quota, as fixed by the Confederate War Department, is 500." The Richmond government eventually armed and uniformed a few companies of Black troops, but as far as is known, they never fired a shot in combat.[49]

Despite the wretched condition of the Confederate military in East and South Florida, Union forces generally remained inside their enclaves along the East Florida coast, St. Johns River, at Cedar Keys, at Fort Myers, and Key West. Several factors contributed to this relative inactivity. Dickison's

---

[47] Proctor, *FL100*, 20 November 1864; Robertson, *Soldiers of Florida*, 307–14; "From Florida," *Augusta [GA] Daily Constitutionalist*, 18 July 1864, 2 (quoting the *Gainesville [FL] Cotton Plant*).

[48] Proctor, *FL100*, 17 December 1864.

[49] *Richmond [VA] Daily Dispatch*, 10 February 1865; Glatthaar, *General Lee's Army*, 452–54; Proctor, *FL100*, 12 December 1864.

victories at Gainesville, Magnolia, and St. Augustine convinced Federal leaders of the danger of sending small units into "Dixie's Land." The defeat at Olustee also remained fresh in their minds, and the bitter animosity between Black and White troops, particularly at Fort Myers, also limited forays from that post. Additionally, the difficulty of fighting in undeveloped regions presented problems that the bluecoats never quite mastered. One soldier, trying to explain the Union's lack of success in the state, wrote,

> It is almost an impossibility to fight the rebels successfully in Florida. In a country covered in extensive forests and immense swamps, having but few good roads, but thoroughly intersected by trails and bridle paths; whose whole surface is known to almost every one of the enemy, but which our force is totally unacquainted, it must continually occur that our forces will meet with loss and defeat while the enemy suffer comparatively little.[50]

FERNANDINA AND NASSAU COUNTY

The situation on Amelia Island had been relatively peaceful since the Federals took it in early 1862. There had been several raids along the St. Marys River and the Nassau County interior, but those remained fairly infrequent events. By December 1864, Pres. Lincoln felt the region along the Northeast Florida coast had been completely restored to its former allegiance. He announced,

> Whereas, by my proclamation of the 19th of April, 1861, it was declared that the ports of certain states, including Norfolk, in the State of Virginia, and Fernandina and Pensacola, in the State of Florida were, for reasons set forth, intended to be placed under blockade: and, whereas said ports were subsequently blockaded accordingly, but having for some time past been in military possession of the United States, it is deemed advisable that they should be opened to domestic and foreign commerce.

[50] "A Letter from the Department of the South," *Rutland [VT] Weekly Herald*, 8 September 1864, 2.

That raising of the embargo came with restrictions against bringing in military supplies of any kind, or anything that would bring harm or discomfort to the area's African American citizens.[51]

Rear Admiral John Dahlgren, commanding the South Atlantic Blockading Squadron, issued strict orders to the blockading ships for dealing with vessels approaching the Southern coast. United States boats, with proper papers, would not be detained or searched. Dahlgren, however, saw potential for problems from foreign ships. He decreed, "Vessels from abroad, however, professing to be bound for Fernandina, even with regular papers, will always be justly liable to suspicion if found in the vicinity of other blockaded ports of this command, and in that case will be subject to seizure and adjudication by the proper courts."[52]

With Amelia Island completely redeemed, Northern newspapers began carrying notifications that land owned by Rebels would be auctioned to the highest bidder. A Wisconsin journal wrote, "U. S. Tax Commissioners will commence [the] sale of [Rebel] Lands at Fernandina on 5 December." Some of the most valuable real property to be sold belonged to former US senator and railroad entrepreneur David Levy Yulee; Mary Martha Reid, wife of a former governor and matron of the Florida hospital in Richmond; and the dwelling of Confederate General Joseph Finegan.[53]

VOLUSIA AND ORANGE COUNTIES

In early October, the Confederates in Volusia County suffered a setback, but their own lack of caution created the problem. The area's Rebels had planned a major recruitment campaign at Enterprise (Volusia's county seat in the 1860s), and apparently failed to take even rudimentary precautions to guard the festivities. A pro-Union newspaper reported,

> Gen. Hatch has recently sent out a force under Col. Noble, which proceeded to Enterprise, a long distance away and captured a whole militia meeting of Volusia County [Rebels], without trouble. The meeting had been advertised and news came to Gen. Hatch.... The

---

[51] "Partial Raising of the Blockade," *Port Royal [SC] Palmetto Herald*, 1 December 1864, 1.

[52] Proctor, *FL100*, 7 December 1864; "Florida: Fernandina," *Milwaukee Sentinel*, 24 November 1864, 1.

[53] "Florida: Fernandina," *Milwaukee Sentinel*, 24 November 1864.

> force started at night, and after a long march succeeded in surrounding the gathering before anyone suspected his [its] presence. Twenty-nine men, forty horses, and some property were secured. Among the prisoners was Col. [William B.] Watson, long known as a successful guerrilla chief, whose specialty has been obtaining cattle from us [the Federals].[54]

Watson had been an early and enthusiastic Rebel. The Georgia native enlisted in Co. H, 2nd Florida Infantry ("St. Augustine Rifles") and served throughout the early battles of the Virginia Army. In late 1863, or early 1864, he transferred to Munnerlyn's Battalion, where he directed the Cow Cavalry Company in Orange and Volusia Counties. A post-war history stated Watson might have escaped capture, but a turncoat pointed the bluecoats to where he had hidden. Sent to a Union prison, Watson remained true to the Confederate cause, finally taking the oath of allegiance in June 1865.[55]

To the east of Enterprise, the New Smyrna area had become a haven for deserters and conscript evaders. With no Rebel forces stationed there, the wild, sparsely settled region became an ideal hiding place for those who wanted no part of the conflict. A Northern newspaper reported that those who had abandoned the Confederate cause "succeeded in eluding their [Rebels'] vigilance by camping out in swamps contiguous to their residences. Those living under the immediate protection of our naval forces were [largely] unmolested."[56]

About the same time that Dickison scored his victory over the Fourth Massachusetts Cavalry at the Big Gum Swamp, a contingent of Rebel horsemen swept through eastern Volusia County gathering up as many deserters as they could find. Federal sources identified the Rebel forces as members of Dickison's unit, stating, "[S]everal parties of rebel guerrillas belonging to Major [Captain] Dickison's cavalry made a concerted move on the inhabitants living near the seaboard, and captured numbers of them, and charging that they had taken the oath of allegiance to the United States government, and had been in the habit of trading with the blockading squadron." The journal

[54] "Capture of a Militia Meeting," *Richmond [VA] Sentinel*, 20 October 1864; Gould, *History of Volusia County*, 89; Hartman and Coles, *BRF*, 1:209.

[55] "A Rebel Raid in Florida," *Philadelphia Dollar Newspaper*, 21 September 1864, 2.

[56] Ibid. Though *Dickison and His Men* lists Mary Elizabeth Dickison as the author, it has long been suspected that J. J. Dickison was the real author.

added, "Great distress prevails among the poorer classes of people in consequence of these raids."[57]

It seems doubtful that Dickison or his men carried out the operation. The newspaper claimed Dickison raided the area "to supply his ranks with conscripts." Florida's "Swamp Fox," however, prided himself in leading a company of true Southern patriots, and he knew that deserters, forced against their will into service, would be unreliable. Also, in two books Dickison makes no mention of this, or similar, operations."[58]

An article in a Union military newspaper seems to confirm that the Volusia County campaign was an *ad hoc* operation. The *Port Royal New South* reported, "Jackson O. Murray, a citizen of Volusia County, Florida, has been condemned to be hanged, on a charge of assisting in hanging John Whitney...on account of said Whitney's loyalty to the United States." However, Murray's name appears nowhere in Dickison's writings, nor in Hartman and Coles's six volume *Biographical Roster of Florida's Confederate and Union Soldiers.*[59]

---

[57] Hartman and Coles' *BRF*, index; "Volusia County, Florida," *Macon [GA] Telegraph*, 16 December 1864, 1 (quoting the *Port Royal [SC] New South*).

[58] Ibid.

[59] Hartman and Coles, *BRF*, index.

Chapter 8

"The Centre Cannot Hold"

(January–May 1865)

THE SITUATION IN 1865

As the New Year dawned, the Southern Confederacy tottered on the brink of collapse. Gen. Robert E. Lee's Army of Northern Virginia, once feared for its slashing offensives, had been cornered within the defenses at Petersburg. Wracked by starvation and decimated by desertions, it little resembled the fearsome force of old. An Alabamian, visiting his former comrades, lamented, "A mere handful remains of the little band [9th Alabama Infantry]; they have been wasted by the storms of battle and by disease, and even the few remaining look weary and worn.... I find our ranks so thinned that when in line of battle along the [breast]works, the boys are scattered eight feet apart. Hardly a skirmish line, yet when the battle comes, they will meet solid [enemy] lines."[1]

The remnants of the Army of Tennessee, which had been uselessly slaughtered on the icy fields at Nashville in December, had barely escaped the Volunteer State. Only desperate rear guard fighting by Gen. Nathan Bedford Forrest's cavalry and Gen. Edward C. Walthall's Mississippians had allowed the survivors of Gen. John B. Hood's army to put the Tennessee River between themselves and the Federal pursuers. A mere shadow of the once proud Confederate "Army of the Heartland," they limped southward into Central Alabama. The Rebels' Western Army had been cursed with mediocre commanders and bad luck, and Gen John B. Hood, the worst of the sorry lot, had been forced to resign when the scope of the defeat at Nashville became known. Pres. Jefferson Davis's only viable command option was Gen. Joseph Johnston. Davis loathed "Old Joe," but the common soldiers loved him. Johnston's first order of business would be to try to gather enough men to halt, or, at least, delay, Gen. William T. Sherman's march through the Carolinas. As the Army of Tennessee limped through Chester, South Carolina, a

[1] Waters and Edmonds, *Small but Spartan Band*, 163–67; Patterson, *Yankee Rebel*, 205–206.

diarist bemoaned, "There they go, the gay and gallant few, doomed, the last gathering of the flowers of Southern pride, to be killed, or worse, to prison...What will Joe Johnston do with them now?"[2]

Gen. Sherman's March to the Sea had been designed to "make Georgia howl," and it accomplished that purpose—in spades. The devastation in Georgia had been great, but its effects had been felt far beyond that state. Marching virtually unopposed pointed out the Confederacy's inability to protect its vital heartland. More importantly, it also temporarily cut the rail lines that transported Florida beef to the South's major armies. An Ohio newspaper reported, "It would appear that the destruction of the Gulf railroad by Sherman struck a staggering blow to Lee and Davis at Richmond...[T]hat road...supplied Lee's army up to the time it was broken with 11,500 head of cattle per week, the cattle from Florida and South Alabama." The writer might have inflated number of cows delivered, but the "March to the Sea" did, indeed, temporarily interrupt the shipment of beeves from the state to the Rebels' major armies.[3]

Despite Union victories elsewhere, the military situation in Florida remained largely stalemated. The Federals had roughly twenty-three hundred troops in East Florida, with a similar number scattered along the coast and through the South Florida interior. As Dr. Coles explains, "While Federal troops in Florida did not heavily outnumber the Confederate defenders, they had superior equipment, and had a much better logistical system. Yet boredom and a sense of isolation kept Federal morale rather low." Perhaps more importantly, Black troops and White Floridians who had joined the Union army, made up most of the state's Yankee force; and a great deal of animosity existed between the races wearing Union blue. That bias, between the Union allies, further helped to hamstring the effectiveness of the state's Yankee soldiers.[4]

By early 1865, despair gripped most pro-Confederate Floridians. Even the most die-hard, like Gov. John Milton, had become convinced that victory was impossible. Despite Rebel victories at Station No. 4, Braddock's Farm, and Natural Bridge, defeat seemed inevitable to most of the state's White citizens: "The Union blockade grew stronger every day, and few southerners

---

[2] Horn, *Army of Tennessee*, 417–24; Chesnut, *Diary from Dixie*, 342.

[3] Sword, *Confederacy's Last Hurrah*, 61–62; "Last Night's Dispatches," *Sandusky [OH] Daily Commercial Register*, 7 January 1865, 3.

[4] Coles, "Far from Fields of Glory," 373–74.

held out any hope for foreign recognition. Unionism and anti-war sentiment, always strong in certain areas of the state, became more visible by early 1865." Military forces, however, remained in the field, and they would continue the fight, and "[w]hen the Confederacy finally collapsed, Florida's defenders were among the last organized troops east of the Mississippi River to capitulate."[5]

Civilians and Dignitaries

Two lines from W. B. Yeats's poem "The Second Coming" best describe the Florida homefront during the final months of the Civil War. The Irish poet wrote, "Things fall apart; the centre cannot hold; / Mere anarchy is loosed upon the world."

A Southern newspaper summed up the situation in Northeast Florida, writing, "Highway murders and robberies are becoming a frequent occurrence in Florida. It is quite unsafe to travel about alone."[6]

Even in Marion County, long considered one of the safest areas in the state's interior, ordinary citizens fell victim to the "anarchy." A journal reported, "A. S. Brown, of Ocala, Florida, was murdered on Friday night, March 24th, while on his way from Ocala to Mrs. Pyles' residence.... Seven bullets entered his body, and he was robbed of his watch." The same article described a second attack, even more grisly than the first. "Recently, a negro girl belonging to Mr. Inman Fisher...entered the room of Mrs. Fisher very early in the morning and with a hatchet...struck or cut Mrs. Fischer on the head in a most horrible manner." The assailant then vanished, "as if into thin air."[7]

The *New York Tribune*, quoting a Southern newspaper, related,

> Highway robbery has become so frequent, and the robbers so bold, that a woman fears to venture out of doors after dark alone, and no man dares to trust himself out at night without first transforming himself into a perambulating arsenal. He must be fully equipped with [a] revolver and bowie knife, and must be prepared to use them on the instant. Men are knocked down and robbed of money, watches, shirt studs, and even clothing.

---

[5] Ibid., 373–75.

[6] Yeats, "The Second Coming"; "From Florida," *Augusta [GA] Chronicle*, 6 April 1865, 3.

[7] "From Florida," *Augusta [GA] Chronicle*, 15 April 1865, 1.

The situation had devolved to the point that daylight robberies had become a commonplace occurrence.[8]

With the military stretched to its limits in holding back the bluecoat invaders, the deserter bands became even bolder in their deprecations. Apparently, while seeking the culprits in the Brown murder, Capt. Howse's Marion County militia unit found "a squad of eight deserters," and hung them on the spot. This, however, was a rare victory in a war the South could no longer win. Her army, in some instances, had lost all hope of victory. The assessment of a Northern newspaper proved accurate. The journal stated, "The [Confederate] county[side] is represented as swarming with Rebel deserters and 'tories,' who are forming into predatory bands, roaming about, murdering, and keeping up a reign of terror." The Federal army and navy even began actively organizing East Florida's Blacks and turncoats into new "cavalry regiments," while expressing shock that "the Confederates permit them to do it without opposition."[9]

During this bleak period, loyal Confederates received news of two deaths that caused deep despondency in East Florida. Madison Starke Perry, a South Carolina native, the state's fourth governor, and former colonel of the 7th Florida Infantry, passed away on 7 March 1865 at Gainesville. He had been an early proponent of establishing a new Southern nation and had engineered Florida's secession from the Union. In a tribute to Perry, a newspaper noted, "He filled positions of trust and honor in Florida, the most honorable of all, as Colonel of the 7th Florida regiment, which he commanded through the Kentucky Campaign." Many in the state lamented his death.[10]

Perry's successor, and frequent political adversary, died a couple of weeks later. It has been accepted for more than a century and a half that Gov. John Milton committed suicide—emotionally devastated by the impending demise of the Southern Confederacy—by shooting himself in the head with a shotgun. (Family members, however, have recently postulated that Milton's death may have been the result of a hunting accident.) The Northern press

---

[8] "Highway Robbery," *New York Tribune*, 11 April 1865, 1.

[9] Ibid.; "A Few of the Troubles of the Rebels," *Washington [DC] Daily National Republican*, 6 March 1865, 2; "By Telegraph," *Macon [GA] Telegraph*, 3 April 1865, 2. See generally, Murphree, "Florida and the Civil War," heading "Florida Unionist."

[10] "Mortuary Notice," *Augusta [GA] Daily Constitutionalist*, 1 April 1865, 3; Voyles, "Remembering former Fla. governor Perry," *Gainesville Sun*, 13 November 2011.

immediately posted unsavory accounts of Milton's checkered past, doing all they could to blacken the reputation of the dead man. Though Gov. Milton made mistakes, he loved his adopted state and the Southern Confederacy; he had done everything in his power to protect the Land of Flowers.[11]

Abraham K. Allison, the president of the state senate, replaced Milton as Florida's sixth governor. A native of Georgia, for many years Allison had been a merchant and lawyer at Apalachicola. Like Milton, he was a devoted Rebel and proved his loyalty to Florida, even taking up arms to protect Tallahassee in the fight at Natural Bridge. Allison's term in office, however, would last only forty-nine days. When he surrendered Tallahassee, Federal authorities imprisoned Allison at Fort Pulaski, near Savannah, for six months.[12]

A New York journal printed an accurate portrayal of the few, and shrinking, regions still under Rebel control in mid-April 1865. The newspaper reported, "The area of the Confederacy has been greatly narrowed of late. Armed opposition to the old flag exist in force in the northern portion of Central North Carolina, in Florida, in Southern Alabama, and in Texas, but the forces are without a head, and acting solely upon the orders of their individual generals." Add the deterioration of social order in the southernmost state, and there is little wonder that a condition of "mere anarchy" enveloped the region.[13]

FERNANDINA

The long occupation of Amelia Island and Fernandina had silenced virtually all Rebel sentiment left in Nassau County. Perhaps nothing provided stronger evidence of the pacification of area residents than an election held in early 1865. A Unionist newspaper enthusiastically reported, "On the 1st instant our people elected a mayor and councilmen by a universal [male only] suffrage vote. The freedmen voted with loyal whites, and nobody was hurt." Adolphus Mott, a friend of Lyman D. Stickney and protégé of Salmon P. Chase, won

---

[11] "Death of Gov. Milton," *Augusta [GA] Daily Constitutionalist*, 8 April 1865, 3; "Suicide of Gov. Milton," *Trenton [NJ] State Gazette*, 2 May 1865, 2.

[12] "Headquarters, Dept. of the South," *Augusta [GA] Daily Constitutionalist*, 25 May 1865. Unreconstructed to the end, Allison led efforts in the Panhandle to deny voting rights to Black people during Reconstruction.

[13] "The Area of the Confederacy," *New York World*, 12 April 1865, 1.

the mayoral vote. A notable postwar Scalawag, Mott had often railed against placing Blacks under the thumb of "prejudiced Southern [White] men."[14]

Another sign of the tranquility of Nassau County had occurred a couple of months previously. A small group of Yankee soldiers and local politicians had ridden southwest, about thirty miles, to Callahan, without encountering the least Confederate resistance. A Unionist reported, "Having reached Callahan, they went to hunt up the [Nassau County] Clerk of the Court and demanded of him the records of his office; stating they wanted them for the purpose of settling titles to property in Fernandina. The Clerk having produced the records, the Yankees took them and quietly left." The need for deed records indicated that Northern investors felt secure enough in the area's stability to begin purchasing property there for postwar habitation and investment. The Federals on Amelia Island, however, still evidenced a reluctance to venture too far up the St. Marys River.[15]

One Federal decision affecting Northeast Florida particularly galled most of the state's White citizens—whether they were Confederate sympathizers or Unionists. Gen. William T. Sherman, before leaving Savannah for his 1865 spring campaign, conceived a plan to both provide home and land for the thousands of Black people who had flocked to the Federal cause during the war. At the same time, it created a buffer zone against Southern Whites along the St. Marys and St. Johns Rivers. An order dated 10 January 1865 decreed, "The islands from Charleston south, [and] the abandoned rice fields along the rivers for thirty miles back from the sea, and the country bordering the St. Johns River, Florida, are [hereby] reserved for the settlement of the negroes made free by the acts of war and the [Emancipation] Proclamation by the President of the United States." Blacks immortalized this edict, believing US government had promised to provide them "Forty acres and a mule."[16]

Southern Whites—even many of those with Unionist sympathies—recoiled at that scheme. Not only did it distribute some of Florida's most valuable and fertile land to African Americans but it would also give the Black men political and financial clout that would help them to achieve true equality. (A Rebel newspaper, however, pointed out the hypocrisy of Sherman's mandate, noting that, "[Most] northern states...have passed laws to prevent

[14] "Fernandina," *Milwaukee Sentinel*, 31 May 1865, 3; Baggett, *Scalawags*, 160.

[15] "From Florida," *Augusta [GA] Chronicle*, 19 February 1865, 2.

[16] "From Savannah," *Augusta [GA] Chronicle*, 4 February 1865, 2.

the [Southern Negroes from] coming within their borders, and competing with the labor of their poor whites.") The same journal, quoting a Quincy, Florida *Dispatch* article, avowed, "We are anxious to see the above experiment tried, particularly in Florida. Lincoln will have to keep an army on the St. Johns, to protect the negroes, or there will be no darkies there a month after they are planted. Just think of it: Florida colonized by runaway and stolen negroes!"[17]

St. Augustine and Picolata

Outwardly, St. Augustine seemed to be almost as "pacified" as Amelia Island. A modern historian observed that in early 1865, "[T]he population of the town increased, [and] other signs of normal order appeared." The citizenry had celebrated the Fourth of July and Thanksgiving; schools were in session; churches held weekly services for the devout; and, the Federal government sold the property of "known rebels" to local Unionists.[18]

These signs of normalcy, however, could not hide the truth from Union officers in Northeast Florida. Confederate irregulars still lurked on the outskirts of the Ancient City, killing and capturing any individual or small group that let down their guard for very long. In January 1865, the provost marshal at St. Augustine received an offer from a civilian, named G. W. Long, to "raise a company for the purpose of repelling and capture rebel scouts [bushwhackers] on the east side of the St. Johns." However, Long had managed to enlist only ten men for that dangerous undertaking. At about that same time, a member of the US Congress objected to spending government funds to repair public buildings in St. Augustine. He reasoned, "I am very much afraid that if we spend money [to upgrade public structures], the guerrillas may interfere with the buildings after we put them in order." The situation for Federals at the Ancient City had become so dangerous that Brig. Gen. E. P. Scammon decreed, "no [military] party [shall] be sent out [of St. Augustine] less than *1000 strong* [emphasis added], except when scouts are sent to

[17] "Grand Yankee Scheme to Colonize Negroes in Florida," *Augusta [GA] Daily Constitutionalist*, 30 March 1865, 1.

[18] Graham, "Home Front," 43; Redd, *St. Augustine in the Civil War*, 43.

ascertain the position of the enemy." Even that precaution failed to deter the determined Rebel brushfighters.[19]

For some time, the Yankees had maintained a sizeable outpost at Picolata on the east bank of the St. Johns River. The Seventeenth Connecticut Volunteers had seen tough fighting at Gettysburg and along the South Carolina coast, and posting them to a military backwater in Northeast Florida represented a kind of reward for their hard service. In early 1864, three companies of the unit were detached to man a garrison at Picolata. A unit historian wrote, "They [the Seventeenth Connecticut] marched and countermarched through the next year, often to little purpose, and [when they] left their outposts along the St. Johns River for brief expeditions into the Confederate interior the men battled heat, disease, insects, boredom and occasional unpleasant surprises at the hands of Confederate guerrillas." To make matters worse, one soldier complained, "[T]he commissary department is in a low state. There is neither sugar nor salt, and of other rations there is not a ten day supply...[The] forage for the horses...is gone."[20]

Capt. J. J. Dickison "had noticed that the Yanks [along the east bank of the St. Johns] seemed to be very casual about their affairs lately." On the evening of 2 February, he began transporting his men and horses across the St. Johns River. The partisan leader had only "one flatboat" to ferry his command across the mile wide waterway, and it took sixteen hours to affect the passage. Dickison likely had a total of 120 men and horses on the east bank by mid-morning. Florida's "Swamp Fox" initially hoped to strike directly at the Union stronghold at Picolata, but quickly realized that an assault—with the force he could muster—would be disastrous.[21]

One of Dickison's men hailed from the region near Picolata, and the trooper quickly fetched his father to their camp. The older man informed the guerrilla leader that a dance had been scheduled in the neighborhood that night and added that a number of Union officers and soldiers would almost

---

[19] *ORA*, ser. 1, vol. 47, pt. 2, 165; "Florida Report of Provost General," 23 March 1865, No. 14356, (G. W. Long); [No headline] *Washington [DC] Daily Globe*, 10 February 1865, 2.

[20] House, "Warren History Background"; Pankenier, *Ridgeland Fights the Civil War*, 34.

[21] M. E. Dickison, *Dickison and His Men*, 114–15; Grzelak, "Five Fateful Days in February 1865."

certainly be in attendance. A Connecticut Yankee gave an account of what happened next:

> The last trap, in which ten or a dozen of these men, with Capt. [Henry] Quinn, were captured, was baited with the feminine gender, and took the shape of a social dance, at the house of a Salina [Solona] family, about 11 miles from St. Augustine. The regimental band was along, and when things were going along nicely, and the 'eyes spoke love to eyes that spoke again,' the 2nd Florida Cavalry pounced upon the whole party, and bore them into captivity. Verily, [female] society is enchanting.

Dickison simply recorded they "captured…about forty men, including four officers, also eighteen horses and one fine ambulance."[22]

FIGHT AT BRADDOCK'S FARM

Dickison's next success—the Battle of Braddock's Farm—seems have begun with a sleazy business deal. Several sources related that, "cotton had been stolen by deserters and sold it to Northern speculators [at St. Augustine]." The merchants had then apparently convinced military leaders at that city to send troops to the East Florida frontier to escort their ill-gotten gains to the Ancient City. Another historian suggests the Braddocks, a "Secesh" family with several sons in the Rebel army, had gathered a large quantity of cotton and cattle at their farm near Dunn's Lake [modern Crescent Lake]. Col. Tilghman assembled the Seventeenth Connecticut and dispatched "several dozen men from various companies [who had] volunteered to retrieve it." Lending credence to the first version is Gen. Scammon's subsequent rebuke in the *Official Records*: "The general also regrets that Colonel Tilghman should have sent out such a small force so far into the country for *purposes not strictly military* [emphasis added].[23]

While at the dance near Picolata, the "Swamp Fox" had learned of the expedition to Braddock's Farm. The Rebel cavalry leader, writing in the third person, penned, "Captain Dickison learned, through a reliable source, that

---

[22] M. E. Dickison, *Dickison and His Men*, 116–17; "The 17th CV's" *New Haven [CT] Columbian Register*, 11 March 1865, 2.

[23] "New from Southern Sources," *Manchester [NH] Weekly Union*, 7 March 1865, 3; Grzelak, "Five Fateful Days in February 1865"; "James Aldridge Braddock"; *ORA*, ser. 1, vol. 45, pt. 2, 165.

Colonel Wilcoxon, with the Seventeenth Connecticut and ten large six-mule wagons, had gone up the old Government Road, in the direction of Volusia County." According to a news article that received wide circulation in the early twentieth century, the source of the intelligence regarding Wilcoxon's expedition came from a brave Southern spy named Lola Sanchez. Even with her father confined at Fort Marion for his pro-Confederate activities and a brother in the Confederate army, the home of the three beautiful Sanchez sisters had become a favorite rendezvous place for unattached Union officers. Lola learned from them about Wilcoxon's operation, and while her sisters, Panchita and Eugenia, continued flirting with the bluecoats, Lola rode through the night to relay the information to Dickison.[24]

Using a reliable tactic of irregular warfare, Dickison and several of his men dressed in blue overcoats they had captured at the dance. In a postwar account, a young witness averred, "They [Dickison's command] continued their march, every few miles meeting deserters on their way to St. Augustine. Gaining all the information desired from them they were sent to the rear as prisoners. All that night, and the next day, they followed the deep ruts of the wagon, loaded down with Sea Island cotton." Early the next morning, two deserters informed the blue clad Rebels that Wilcoxon was at the Braddock Farm, about two miles south of their location.[25]

Dickison's Rebel guerrillas arrived at the Union encampment shortly after dawn on the morning of 5 February. One of Braddock's sons rode with Co. H, and with young Braddock's help, Dickison arranged his troopers carefully. He directed Sgt. William Cox, with ten men, to block the government road along the north bank of a stream (likely Pallicer Creek) that the wagon train would have to cross to reach St. Augustine. After carefully arranging the rest of his force—about forty-five men—Dickison waited at the foot of a small hill.[26]

---

[24] M. E. Dickison, *Dickison and His Men*, 116–18; "Fair Southern Spies," *New York Daily People*, 11 August 1905, 3.

[25] M. E. Dickison, *Dickison and His Men*, 116–18; Grzelak, "Five Fateful Days in February 1865"; "The Battle of Braddock's Farm," *Palatka [FL] News and Advertiser*, 10 March 1911, 5. The Braddock farm was located near the modern municipality of Crescent City.

[26] Grzelak, "Five Fateful Days in February 1865"; M. E. Dickison, *Dickison and His Men*, 120–25.

The engagement began when some of Cox's men prematurely opened fire on the approaching wagon train. Spurring their horses and screeching the Rebel yell, Dickison's cavalry dashed into the surprised bluecoats. A few of the Unionists returned fire, but most surrendered almost immediately. A Yankee officer related,

> The [Union] force was about on its return home, when it was attacked by about 200 hundred [*sic*] of Dixon [Dickison's] mounted rifles. The attack was sudden and unexpected. They [surprise attacks] were easily made so in Fla., which was, pretty much one [continuous] pine wood. A summons to surrender was unheeded by Col. [Albert H.] Wilcoxon, and fire [from both groups] opened. Seeing no hope to escape Col. Wilcoxon and Adjt. [Henry] Chatfield attempted to cut their way thro[ugh] the enemy. Adjt. Chatfield was instantly killed and Col. Wilcoxon shot thro[ugh] the shoulder of which he afterwards died at Tallahassee.[27]

Dickison spurred his horse toward Wilcoxon, describing the encounter like medieval knights engaged in a jousting tournament. Again writing in the third person, he recorded,

> Captain Dickison...seeing this officer approaching...rode out to meet him, and demanded a surrender. Driven to desperation, he [Wilcoxon] drew his sword and made a furious charge at the captain, who fired, the shot taking effect in his left side. As their horses were moving rapidly, they passed each other. Captain Dickison quickly turned and soon gained upon his adversary, whose glittering sword flashed defiance. Again the captain fired with sure aim. The saber strokes falling heavy and fast. One more shot, and his antagonist fell.

A Federal soldier stated that Adjt. Chatfield "encountered one of the [Rebel] lieutenants and a fierce hand to hand contest ensued, at the end of which he, too, fell [wounded]...and a [graycoated] soldier brained him where he lay."[28]

After his severe wounding, Wilcoxon flashed the Masonic order's distress sign; Dickison and the unit surgeon, both Masons, immediately

---

[27] Ibid.

[28] Ibid.

responded to the signal. Though the Union officer had been desperately wounded, the two Confederates made him as comfortable as possible. The courageous Federal officer succumbed to his wounds a couple of weeks later at Tallahassee. (Some Federals suspected the Rebels had killed the Connecticut officer, but that seems highly unlikely, considering Dickison's admiration for and fraternal connection to the Union officer. After receiving a plea from Wilcoxon's wife, the "Swamp Fox" of Florida bent the rules of war and returned the Connecticut officer's sword, which had been a gift from his Masonic brothers.)[29]

Dickison and his men moved quickly, crossing the St. Johns that very night. He reported "taking 51 prisoners (including a Lt. Col. & 2 captains), killing four men, also [capturing] 18 deserters and tories, 10 wagons & teams with seed cotton (about 9,000 pounds), and a number of small arms and horses." Dickison wrote that he "re-crossed the river…without the loss of a man."[30]

SOUTH FLORIDA AND CATTLE COUNTRY

South Florida remained relatively quiet during the final months of the war. The East Gulf Blockading Squadron continued doing good service, keeping many of the blockade-runners bottled up in the rivers and bays of the peninsula's west coast. Confederate defeats at Charleston and Wilmington allowed the Union to shift men and ships to Key West and the southern part of Florida. A sailor wrote, "Our squadron has been reinforced by several steamers, which has been relieved from the Wilmington blockade. Admiral [Cornelius] Stribling has made such disposition of them as, I think, will soon stop blockade-running between Havana and the coast of Florida."[31]

In the middle of February, Maj. William Footman, commanding the Cow Cavalry in Southern Florida, with a force of 150–250 men, made an attempt to capture the Union garrison at Fort Myers. Why Footman undertook this operation remains something of a mystery. The Cow Cavalry's primary objective had always been to move cattle to the northern part of the state (for shipment to the Confederate armies of Lee and Johnston); the Yankee forces had done little to interfere with Confederate operations since the middle of the previous year. In addition, rumors had spread throughout the

[29] Ibid.

[30] Ibid.

[31] "Florida," *Philadelphia Press*, 14 March 1865, 1.

region that the Federals intended to soon abandon the bastion. Rodney Dillon Jr., one of the early historians of the Civil War in South Florida, suggests the action was primarily symbolic. "[S]outh Florida Confederates," he wrote, "remained determined to continue their resistance and to eliminate Fort Myers, a major source and symbol of Union power in southern Florida."[32]

Footman's Rebels scored a complete success in their initial action. They captured the outer line of Yankee pickets—"a corporal and three privates from the Second Florida (Union) Cavalry"—a few miles from Fort Myers. This apparently occurred at Billy Creek, and the Confederates still held the element of surprise.[33]

As the Rebels advanced toward Fort Myers, a downpour began, which significantly affected the outcome of the battle. Lt. Francis C. M. Boggess, a reluctant Confederate and veteran of four wars, reported, "On the night that our anticipated attack was to be made it rained until the water was knee deep over the entire country." Not only did it drench the region around the fort, but also the ammunition for the attacker's muskets, side arms, and cannon shells. In addition, the downpour threw the Cow Cavalry well behind their schedule, and they did not reach Fort Myers until almost noon.[34]

The plan for a surprise attack on the Union bastion quickly deteriorated into a comedy of errors. Footman sent a demand for the Yankees to surrender, which the bluecoats promptly rejected. The Confederates' lone cannon opened on Fort Myers, and the two artillery pieces from the stockade responded. The Union fire proved so accurate that Footman's men were forced to move their cannon at least three times during the engagement. Lt. William McCullough, who had returned from his banishment to Cedar Keys, led a Union cavalry company outside the walls and formed a skirmish line around the fort. Another small Yankee mounted unit dashed out of the citadel to bring in the Federals' cattle and horses.[35]

That night the Confederates rounded up a few beeves, which they butchered, cooked, and ate. The next morning the Rebels held a council of war. The men agreed to stay and continue the fight, but Footman realized

---

[32] Dillon Jr., "Battle of Fort Myers," 30; R. Taylor, "Cow Cavalry," 211–12; Bamford and VanLandingham, "Now Just Hold Your Cow Cavalry Horses!"

[33] Dillon Jr., "The Battle of Fort Myers," 31.

[34] Ibid.; Boggess, *Veteran of Four Wars*, 67–70.

[35] Dillon Jr., "Battle of Fort Myers," 31–32; Boggess, *Veteran of Four Wars*, 67–70.

the futility of prolonging the "siege." The Confederates began their withdrawal, and Union scouts trailed them, reporting that Footman's men were "in full retreat to[ward] Fort Thompson."[36]

The Battle of Fort Myers accomplished almost nothing. Dillon concluded, "The military situation in southwest Florida following the attack on Fort Myers was little different than it had been before the fight." The Yankees lost one man killed and eleven captured, and despite inflated Federal accounts, the Confederates seem to have had no casualties. Thereafter, the bluecoats accelerated the withdrawal of their forces from Fort Myers, and the Cow Cavalry, continued sending cattle out of the state.[37]

If Lt. Boggess is any indication, Maj. Footman lost both the confidence and respect of his men during the fight at Fort Myers. Boggess declared his commander "a complete failure." Maj. Sam Jones apparently concurred. He requested permission to advance J. J. Dickison to the rank of colonel and put him in charge of operations in South Florida. Gen. W. J. Hardee, commanding the Department of South Carolina, Georgia, and Florida, assented to the change, but the war ended before the reassignment could be finalized.[38]

Spring marked the beginning of the new cattle-driving season in Florida, but news of the surrender of Gen. Lee's Army reached the state by mid-April. "It was obvious that further bloodshed was pointless," Boggess wrote, but it took until 5 June before the members of Munnerlyn's Battalion received their paroles at Bayport. The Yankees went out of their way to humiliate the Rebels, assigning Black troops to accept their surrenders. Lt. Boggess said he resented giving his parole to an African American, but most of the men had had their fill of war. He swallowed his pride, signed his parole, and headed home to his wife and family.[39]

CEDAR KEYS AND LEVY COUNTY

By the first week of February 1865, the situation at Cedar Keys had dramatically improved. Reports of deaths and illness due to the tainted provisions had abated, and the health of refugees and residents seemed to be getting better. That week also marked the return of Maj. Edmund C. Weeks. A New

---

[36] Ibid.

[37] Dillon Jr., "Battle of Fort Myers," 32–33; Bamford and VanLandingham, "Now Just Hold Your Cow Cavalry Horses!"

[38] Boggess, *Veteran of Four Wars*, 72–73; R. Taylor, "Cow Cavalry," 213–14.

[39] Boggess, *Veteran of Four Wars*, 74; R. Taylor, "Cow Cavalry," 214.

York newspaper related, "Major Weeks, it may be remembered was the officer who was recently tried here [Key West] for shooting a sentry.... [He has been] acquitted and reinstated in his command." (The writer failed to mention that Weeks had only obtained the "not guilty" verdict by the use of a reprehensible technicality.) Weeks, however, seemingly still burned with a desire for advancement in rank and martial glory.[40]

Weeks did not wait long to carry the war to the Levy County Confederates. On 8 February, he left the protection of his gunboats and advanced into the countryside. Weeks took with him 186 men of the Second Florida Cavalry (US) and 200 men of the Second United States Colored Troops. Maj. Benjamin C. Lincoln commanded the Black soldiers. They carried with them "knapsacks, haversacks, canteens, and eight days' rations." According to historian Gary Loderhose, Weeks planned to accomplish what Gen. Truman Seymour had failed to achieve during the Olustee Campaign. He wrote, "Weeks' plan included invading the Confederate interior, moving through Levy County to Levyville, then advancing toward Newnansville and Clay Landing. The final objective being [the destruction of] the railroad bridging the Suwannee River near Columbus."[41]

Weeks quickly divided his force. Maj. Lincoln's contingent moved north along the Suwannee to Clay's Landing. Only a handful of Rebel home guards patrolled that section of the county. Weeks wrote, "The force under Major Lincoln surprised, but did not succeed in capturing the [Confederate] company at Clay Landing; they made their escape across the river in boats. He [Lincoln] destroyed a good amount of commissary stores and other Government supplies." A USCT sergeant recounted, "[We] were attacked by the Rebels at a place called Day [Clay's] Landing. In this engagement we lost but one [killed]." After that brief skirmish, the Black troops marched across country toward a community bearing the ill-considered name of Sodom.[42]

Weeks's column, meanwhile, had marched northeast to the Yearty Farm where they surprised a small Rebel picket post. They captured three of the graycoats and four horses. Weeks then pushed on toward Levyville, where

[40] "Our Key West Correspondence," *New York Herald*, 13 March 1865, 5.

[41] *ORA*, ser. 1, vol. 49, pt. 1, 40–41; Sgt. H. C. Jones, "From the Regiments," *New York Weekly Anglo-African*, 1 April 1865; Loderhose, *Way down upon the Suwannee River*, 114.

[42] *ORA*, ser. 1, vol. 49, pt. 1, 40–41; Sgt. H. C. Jones, "From the Regiments," *New York Weekly Anglo-African*, 1 April 1865.

they collected about fifty "contrabands." Many of the slaves they took had no desire to leave, and Weeks felt "obliged to detach most of one company to guard prisoners and [the] contrabands." Lincoln and his Black troops soon rejoined Weeks's contingent at Levyville, and the commander decided to return to Cedar Keys. Why they chose not to proceed north, to burn the Suwannee River railroad bridge, is unknown.[43]

As the Federals began their return trip, a small unit of Confederate cavalry, numbering about twenty men, attacked the rear of the Yankee line. The horsemen almost certainly belonged to Capt. Edward J. Lutterloh's Cow Cavalry Company. Lutterloh had already sent out a call for assistance to the adjoining counties, while the small Cow Cavalry unit trailed the bluecoats, wounding two of the enemy, before the Unionists reached Station No. Four.[44]

Lutterloh's pleas for help soon bore fruit, and a few militia units began arriving. These included Capt. Phillip B. H. Dudley's Alachua County Home Guard, Capt. Thomas L. King's unit from Levy and Marion counties, Capt. J. W. Price's Alachua County company, and a leader named (possibly Robert) Waterson, whose rank and place of origin is unknown. Of greater military assistance, Capt. J. J. Dickison and his men arrived on the night of 12 February. According to Dickison, when he arrived, he found "Captain Lutterloh, with eighteen men, from the outpost [Levy County]; the militia, numbering thirty-seven men, under Captains King, Dudley, Price, and Waterson, making our entire force one hundred and sixty men, including our artillery." Lt. T. J. Bruton had accompanied the cavalry with a single twelve-pounder [artillery piece] and a handful of cannoneers.[45]

During the hasty move to assist Lutterloh, Dickison's men and horses had been worn to a frazzle. In the last two weeks they had campaigned along both sides of the St. Johns River, and the guerrilla leader described Co. H as "almost broken down." When the reinforcements arrived at Levyville, Florida's "Swamp Fox" decided to rest his men and mounts. Dickison wrote, "Just before sundown they [the enemy] reached No. 4, near Cedar Keys. We

[43] *ORA*, ser. 1, vol. 49, pt. 1, 40–41; Levy County Archives Committee, "Yearty Family," 2–12.

[44] Levy County Archives Committee, "Yearty Family," 2–12; Fishburne Jr., *The Cedar Keys in the 19th Century*, 70–75.

[45] *ORA*, ser. 1, vol. 49, pt. 1, 42–43; M. E. Dickison, *Dickison and His Men*, 137–38.

[Dickison's men] were about four miles in the rear. Night coming on, a halt was ordered, and a very strong picket put out, [so] that it would be impossible for the enemy to surprise me." The men and their "critters" needed the extra rest. Gen. William Miller was also leading a large group of reinforcements from the Panhandle region, and the respite would allow Miller a chance to close the gap toward Levy County.[46]

Two very different versions emerged regarding the events of 13 February 1865. Maj. Weeks, of course, boasted that he had scored a major victory over an "overwhelming" Rebel force. He got all the cattle and contrabands to safety at Cedar Keys, he claimed, returning in time to save the day when his Black troops disintegrated under Confederate pressure. He wrote that he led a charge that drove the graycoats off the field, in great confusion, even though Gen. Miller had reinforced Dickison's Rebels. Weeks's official report averred,

> At 7 [o'clock]...[I] heard firing at Station Four. Returned there as soon as possible; found our men flying in all directions; [and] left an officer to halt and bring them up.... At this time Captain Pease with about forty men...charged the enemy who were making an attack on our camp. The enemy, from 250 to 300 strong, with two pieces of artillery, commenced giving way. We took the bridge, and as soon as possible after crossing I deployed my men on the right and left of the road as skirmishers; drove the enemy gradually back until they broke and took flight.

He related he lost five killed, thirteen wounded, and three captured.[47]

Dickison reported he found a heavy force of the enemy behind the rail bed, "the high embankment making for them a good breastworks." The Federals fired on Dickison's skirmishers, but were soon driven into the railroad by a combination of rifle and artillery fire. The guerrilla leader wrote, "The enemy, in their stronghold, gave way...and in a few minutes we held the road. The fight then became general, and our artillery shelling them at a furious rate. They [the enemy] would give way, but [would] rally again and again, and renew the attack." The Negro troops, early in the engagement, had foolishly raised the cry of "No Quarter," and they paid dearly for that bravado.

---

[46] *ORA*, ser. 1, vol. 49, pt. 1, 42–43; M. E. Dickison, *Dickison and His Men*, 137–38; Evans, ed. (and J. J. Dickison), *Confederate Military History* (Florida), 11:206–207.

[47] *ORA*, ser. 1, vol. 49, pt. 1, 41.

"No quarter" indicated that they intended to show no mercy to the enemy—and expected the same from their opponents. A member of Lutterloh's Cow Cavalry, writing postwar, stated, "[A]bout 200 Negro troops with white officers marched back over the railroad...[T[he unsuspecting troops were surprised and overwhelmed by the [Confederate] troops lying in wait for them...and under [our] onslaught, soon broke ranks, and ran into the black marsh grass, where they were chased and dispatched with expediency by the vengeful Southern cavalry."[48]

Near darkness, the Rebels ran out of ammunition for both their arms and artillery, and fell back about a half a mile, where they awaited Gen. Miller's reinforcements and a fresh supply of powder and shot. Miller arrived during the night, but the Federals had already taken advantage of the lull in the fighting to retreat to Cedar Keys. The Cow Cavalryman added, "The slaves, horses, and several hundred head of cattle, with other stolen property, were recovered and returned to [their] owners." Dickison reported he had no deaths, but five men severely wounded.[49]

Despite Weeks's heroic account of the battle, apparently few at the time believed his narrative. A headline regarding the fight at Bridge No. Four, in the *New York Herald*, described the operation as a "Disastrous Raid of Major Weeks.'" One of the Black sergeants in the fight, writing to the *New York Anglo-African*, bluntly wrote, "The sun rose clear and beautiful, and *many* [emphasis added] who saw it rise never witnessed its setting.... Our men [USCT] stood up bravely, although the thing was *managed poorly by those in command* [emphasis added].[50]

A few weeks later, Maj. Weeks and his soldiers participated in the fight, near Tallahassee, at Natural Bridge. Upon their return to Cedar Keys, however, Weeks seemed content to remain at the island city. The Rebels controlled the rest of Levy County, and an African American soldier from Key West, on a raid along the Wekiva River, spoke to a "Secesh" lieutenant under a flag of truce in early April. The young Confederate officer expressed himself

---

[48] M. E. Dickison, *Dickison and His Men*, 138; Sgt. H. C. Jones, "From the Regiments," *New York Weekly Anglo-African*, 1 April 1865; Starkey, "1st Florida Special Cavalry."

[49] M. E. Dickison, *Dickison and His Men*, 138–39.

[50] "Disastrous Raid of Major Weeks," *New York Herald*, 13 March 1865, 5; ; Sgt. H. C. Jones, "From the Regiments," *New York Weekly Anglo-African*, 1 April 1865.

tired of the war but claimed that voicing any sympathy for the Union cause was a death sentence in the backwoods of Levy County.[51]

JACKSONVILLE AND THE ST. JOHNS RIVER

With horrific crimes on the rise, and the news from Virginia and the Carolinas getting gloomier by the day, many Floridians sought a safe haven in the Union-occupied enclaves. Citizens of Jacksonville considered themselves relatively safe from direct attack, but their sanctuary was far from secure. Dr. Schafer observed, "By the end of 1864 it was clear that the major infantry and artillery engagements in northeast Florida had ended, yet alarms and raids were frequent enough in early 1865 to remind Jacksonville residents that the shooting had not ended." Enemy territory began just beyond the range of the Federal gunboats' cannons; Gen. Scammon's recent decree to avoid sending patrols into the Rebel frontier "less than 1,000 strong," revealed just how tenuous was their position.[52]

All was not "gloom and doom" in the River City. For example, on 26 February the Yankees fired a one hundred-gun salute in honor of their army's capture of Charleston, South Carolina. The Northern troops at the River City also made occasional raids into enemy territory. A news journal reported, "[I]n early March, Yankee raiding party made a successful foray to Newnansville, bring[ing] back twenty-two slaves and as many captured horses and mules." Another incursion, deep into Rebel territory, included five hundred members of the Thirty-fourth USCT, commanded by Col. W. W. Marple, which supported the successful effort by a "crew of wreckers" to salvage the vessel *St. Mary's*, scuttled by C. J. and Charles Hemming in Dunn's Creek in early 1862.[53]

Florida guerrillas, however, found a way to use the increasing number of those taking refuge in the River City to their advantage. Early in February, a family from Baldwin appealed to the Union commander at Jacksonville to send soldiers to protect them while they loaded their necessities for a move to the River City. A Northern newspaper related,

---

[51] "Naval Correspondence," *New York Weekly Anglo-African*, 20 May 1865, 3.

[52] Schafer, *Thunder on the River*, 234; *ORA*, ser. 1, vol. 47, pt. 2, 165.

[53] Schafer, *Thunder on the River*, 234–35; "Florida Items," *Augusta [GA] Daily Constitutionalist*, 31 March 1865, 2.

> On Thursday forty men [of the Seventy-fifth Ohio], in command of a captain, left Jacksonville and proceeded about fifteen miles in the direction of Baldwin, their object being to assist a resident family in removing themselves to within our lines.... [W]hile engaged in loading the wagons [they] were surrounded by 400 mounted rebels. Our men fled into the woods, where they were closely followed. The result was that only fifteen of our men...succeeded in getting back to Union lines.

That type of ambush reminded the bluecoats of the danger encountered by small squads in leaving the safety of Jacksonville's defenses.[54]

RAID ON MARION COUNTY

Marion County had provided the Confederacy with some of its finest guerrilla leaders including J. J. Dickison, W. E. Chambers, and J. W. Pearson. In 1860, it had a robust economy based on agriculture, and a population of 3,294 Whites, 5,314 Black slaves, and one free man of color. Many of the region's leading citizens originally hailed from South Carolina. A newspaper editor recalled that in the prewar era, "[We would] meet caravan after caravan daily, and [we] asked where they came from and where were they bound, and nine times out of ten the answer would be, 'From South Carolina, and bound for Marion County, Florida.'" Not surprisingly, they brought with them the Palmetto State's political and social mores. By fall 1860, secessionist meetings had become commonplace; in December, Ocala residents raised a flag on the courthouse square with a single blue star on a white field bearing the inscription, "Leave Us Alone." When the call to arms finally came, the county furnished the Confederacy twelve companies. At least 90 percent of its eligible White citizens joined the military. By the last four months of the war, enemy gunboats seem to have penetrated some of rivers along Marion's sparsely settled eastern border, but as far as is known, by 1865 the bluecoats had never set foot in the more populous area.[55]

---

[54] "Sherman's Army," *Columbus [OH] Daily Ohio Statesman*, 9 February 1865, 4.

[55] Ott and Chazal, *Ocali Country*, 75–79; Riley, "Civil War Years," *Ocala Star Banner*, 1 January 2003; E. M. Graham, "East Florida—Its Progress and Improvement," *Cedar Keys Telegraph*, 3 March 1860, author's collection, Rome, GA.

A Georgia soldier, briefly stationed in western Marion (following the fight at Olustee) left a vibrant description of the Rebel region. "We hear of the Yankees every day," wrote Thadeus Oliver, (composer of "All Quiet along the Potomac Tonight"), "but they are not encamped at any place west of the St. Johns, but only come out in small foraging parties…and retreat hurriedly across the river to their encampment." Oliver also compared Marion favorably to Alachua County. "These people are better than [those] around Gainesville, and cannot tolerate a Yankee or deserter…. Here they hunt them with dogs."[56]

The first known action in Marion County likely occurred at Salt Springs, in the modern Ocala National Forest. Located almost fifty miles from the nearest salt water, citizens of Marion County had apparently made use of that natural resource to provide their salt supply. The *New York Herald*, quoting a Rebel news item, reported, "The enemy [Federals] landed a few days ago [in January 1865] from their ships, smashed up the salt works in the vicinity of the Oklawaha River, Florida, took [the] kettles, killed eight mules, carried off negroes and considerable property." They also "captured Thomas Munroe…and retreated to their vessels." (Munroe, one of the guards at the salt works, appears to have been a young teen.)[57]

As so often happened in Civil War Florida, three separate—wildly differing—accounts of the next raid into Marion County found their ways into the newspapers and *Official Records*. The first Union version declares,

> During March 1865, Sgt. Maj. Henry James of the 3d USCI [United States Colored Infantry] led a raid up the St. Johns River. Twenty-four men of his own regiments and the 34th USCI, one soldier from a White regiment, and 7 civilians (Colored) made up the party. Near Welaka, they hid their boats in a swamp and struck westward about forty-five miles until they came near Ocala. On the way, besides burning a sugar mill and a distillery, they managed to take four White prisoners and set free ninety-one Black Floridians. Returning [toward St. Augustine], they drove off more than fifty Confederate cavalry, who attacked them near the city gates, killing

[56] "Tories in Gainesville," *Ocala [FL] Star-Banner*, 2 June 1964 (reprint of a Civil War letter in possession of an Oliver descendant).

[57] "Raid on Florida Salt Works," *New York Herald*, 29 January 1865, 5.

> at least thirty Southern horsemen. The raiding party's loss during the five-day expedition was two killed and four wounded.

Gen. Gilmore praised the Holley and Marshall Plantations raid as "an example worthy of emulation."[58]

The second narrative seems to have been constructed from rumors and incomplete news reports. A Pennsylvania newspaper begins by turning Sgt. James into "Sgt. Benn." After describing the raid, the journal averred,

> During their trip, while at Lake Church Hill [Churchill], they were attacked by a force of rebels, three times their number, and finding themselves surrounded, thought it prudent to surrender. Sergeant Benn, seeing a rebel approaching, told him they would surrender his forces if they would be treated as prisoners of war. The rebel assented; but when, by this ruse he had approached very near, he took deliberate aim and shot the Sergeant dead. A general charge upon the colored force followed. They all took cover, and one brave fellow sung out, 'Remember Fort Pillow!' Every man took careful aim, and the rather unexpected result was that the whole rebel force was repulsed, with slaughter.

According to this account, the Federal force had only two wounded and Sgt. Benn killed. Confederate casualties amounted to twenty-seven dead, four wounded, and "a great many captured."[59]

The narrative currently accepted by most historians, and the public, tells a much different story from the two previous chronicles. Sgt. Maj. Henry James led a force of about fifty men into Marion County, where they destroyed the Holley and Marshall plantations, freed a great many slaves, and collected a large amount of livestock. According to the wife of a Confederate captain [serving in Virginia], "The news [of a raid by Negro soldiers] having spread like wildfire, they were met by a squad, mostly of old men and boys,

---

[58] Dobak, *Freedom by the Sword*, 87; *ORA*, ser. 1, vol. 47, pt. 3, 190. Three weeks later Dickison surrendered 117 men, making the loss of "30 men," ludicrous, unless the guerrilla leader possessed the power to raise the dead.

[59] "A Successful Raid," *Philadelphia Illustrated New Age*, 12 April 1865, 2. Fort Pillow, north of Memphis, Tennessee, had been the site of a "massacre" of surrendered Black troops by Gen. Nathan Bedford Forrest's Rebel cavalry. Historians still disagree as to whether the "slaughter" occurred during the battle or after their surrender.

hastily gathered from the vicinity of Ocala, commanded by General [Robert] Bullock, who was home on wounded furlough." Capt. John Howse's home guard cavalry also joined Bullock's group. The militia units, often mockingly referred to as "Armies of Last Resort," generally had a poor reputation in combat, but in this case, the local militia put up a stubborn fight. They managed to rescue Frank Holley, one of Marion County's earliest settlers, who had six sons fighting in the Rebel army. "The raiders afterward made a stand and General Bullock's party gave up pursuit." The bluecoats killed two of the home guard unit—a one-armed ex-Rebel soldier named Morrison, and an "almost totally blind" resident named Henry Higgins. John L. Mathews also received a severe wound in the skirmish.[60]

To prevent pursuit, Sgt. James burned the Sharpe's Ferry Bridge over the Oklawaha River, and it turned out to be a very wise precaution. A local historian noted, "Capt. J. J. Dickison, the vigilant protector of Florida homes...receiving information of their raid, and that they had retreated in the direction of the St. Johns River, hastily followed with the view of overtaking, or intercepting them." Discovering the bridge at Sharpe's Ferry had been destroyed, and not wanting to waste time bringing up boats, Dickison rapidly moved to the northeast, finally crossing the St. Johns at Horse Landing.[61]

Though almost a day behind the raiders, Dickison's troop gradually closed the gap on Sgt. Maj. James's heavily laden outfit. Within two miles of the Ancient City, the Florida guerrillas overtook the bluecoats. In order to save their lives, James left most of his spoils to the Rebels, and escaped into the St. Augustine walls. Dickison modestly (a word not often associated with

---

[60] Fannie R. Gary, "The Marshall Swamp Raid: A Thrilling Episode of the Civil War," *Ocala [FL] Banner*, 19 May1905; Bradshaw, "Armies of Last Resort," 30–35; Evans, ed. (and J. J. Dickison), *Confederate Military History* (Florida), 11:133–34; "Historic Markers Across Florida," [**see note in bib**]; David Cook, "Federal troops raid plantation in invasion of Marion County," *Ocala [FL] Star-Banner*, author's collection, Rome, GA. Gen. Robert Bullock, from Marion County, had been seriously wounded at the Battle of the Cedars, at Murfreesboro, Tennessee, in December 1864.

[61] Gary, "The Marshall Swamp Raid"; Evans, ed. (and J. J. Dickison), *Confederate Military History* (Florida), 11:133–34.

him) reported that he took "twenty-four negroes, several deserters, wagons, mules, etc.…recaptured from the enemy within a mile of St. Augustine."[62]

Possibly the last known irregular military action in Northeast Florida occurred near St. Augustine in late April. Capt. Dickison kept a small squad of three to five stationed in St. Johns County to provide him with news of Union activities in the area. A small squad of Federal mail carriers, traveling from Jacksonville to the Ancient City, disappeared. Apparently, the Confederates killed the messengers and took their mail. And, with that sad and unnecessary bloodletting, the guerrilla war in East Florida sputtered to a close.[63]

The Final Days

Like virtually everything about the war in Florida, the end came in "fits and starts." On 17 April 1865, Gen. Joseph E. Johnston surrendered the remnants of the Army of Tennessee to Gen. William T. Sherman. In accordance with Pres. Lincoln's previous directives, Sherman gave very generous terms to the Rebels in an attempt to facilitate a smooth transition back to US citizenship. Pres. Andrew Johnson, who succeeded the slain president, considered Sherman's conditions too lenient, and ordered his general to resume hostilities. Ignoring that disastrous command, the Union general accepted Johnston's surrender on 26 April 1865 at Durham's Station, North Carolina. Since that Rebel officer's command included Southern troops in North and South Carolina, Georgia, and Florida, Sherman included those soldiers in the compact of capitulation. Fewer than thirty thousand men remained of the Army of Tennessee, but the surrender agreement included a total of more than eighty thousand soldiers. Unfortunately, no one thought to notify the graycoats in Florida.[64]

Near dusk, on the evening of 24 April 1865, a Confederate soldier bearing a white flag appeared at the outskirts of Jacksonville and requested to deliver a letter to Gen. Israel Vodges. Vodges, a Pennsylvania native and graduate of West Point, had seen considerable action while stationed along the South Carolina coast. He now commanded Union forces in the Department of Florida. Maj. Gen. Samuel Jones, who had previously been urging his

---

[62] *ORA*, ser. 1, vol. 47, pt. 3, 190; M. E. Dickison, *Dickison and His Men*, 209–10 and 253–55.

[63] "From Florida," *Augusta [GA] Daily Constitutionalist*, 2.

[64] "Johnston Surrenders at Bennett Place" [**see note in bib**]; Coles, "Far from Fields of Glory," 378–80.

troops to press the war with greater zeal, now wanted only to conclude the conflict. The following day the two generals agreed on a temporary cease-fire. A newspaper noted that the "meeting between the two generals was courteous, the conference lasting several hours." Throughout the discussion, Vodges stressed that Southerners must immediately recognize that all Black people were now free.[65]

While Vodges and Jones hammered out an agreement to bring Florida's war to a close, the City of Tallahassee fell to Brig. Gen. Edward M. McCook. That Union officer was part of Maj. Gen. James H. Wilson's expeditionary force. Wilson's horse soldiers had swept through western Alabama and captured both Columbus and Macon, Georgia. Upon hearing that Tallahassee remained in Rebel hands, Wilson sent McCook to take the state's capital city. The Union horsemen entered Tallahassee without a shot being fired. Gen. Jones immediately surrendered the city and its public stores. A Northern newspaper related, "Gen. McCook's method was such as to disarm bitterness. He was cordially welcomed by the citizens, who assured him there was no longer the least disposition among the people of Florida to act in opposition to the wishes of the government, and reconciliation and reconstruction at the earliest possible were most earnestly desired."[66]

Gen. Vodges, who had been given control of the entire State of Florida, deeply resented McCook "stealing his thunder," but there was little he could do about the situation. McCook, however, left Tallahassee about a week later, and Vodges "sent an officer to Tallahassee to assume command." The Third USCT joined the soldiers sent to Florida's capital city to keep the peace, and the presence of the African American soldiers caused some animosity—even among some of the Yankee combatants who had never served with Black troops. Tallahassee's White citizens especially resented presence of Black warriors, but they had few options except to hold their tongues and pretend to be pleased.[67]

Vodges designated several towns throughout the peninsula as places to accept the paroles of Rebel soldiers. Tallahassee, Jacksonville, St. Augustine,

---

[65] "Florida," *New York Herald*, 4 May 1865, 5; *ORA*, ser. 1, vol. 47, pt. 1, 166–67.

[66] "The Surrender of the Rebels in Florida," *Boston Daily Advertiser*, 16 June 1865,1; Coles, "Far from Fields of Glory," 381–84. See generally, J. Jones and Rogers, "The Surrender of Tallahassee," 103–10.

[67] Coles, "Far from Fields of Glory," 384–87.

and Fernandina were chosen, of course, but so were smaller towns such as Madison, Lake City, Tampa, Waldo, Bronson, and Bayport. Though many die-hard Rebels put it off as long as possible, "by the last week in May 1865, all recognized Confederate units in Florida had capitulated."[68]

During late May, J. J. Dickison assembled his company at Waldo for the last time. There had been considerable discussion among his troops urging their commander to lead his unit to the Trans-Mississippi region to join the army of Gen. Edmund Kirby Smith. Dickison, however, seems to have realized the futility of that course of action. In the final address to his soldiers, Dickison said, "Your noble deeds of heroism have crowned you with glory and renown.... To separate myself from such men, after the relationship, which has existed between us is a severe trial...I am proud to know we are not whipped—only overpowered. We stand firm, unshaken, united." His soldiers left Waldo, and entered a new world, where their slaves—the primary source of the South's prewar wealth—had been set free, and their enemy reigned supreme.[69]

BENJAMIN AND BRECKINRIDGE

One group of Confederates—Jefferson Davis and his cabinet—could neither be paroled nor forgiven. A modern writer explained. "There was a good reason for their [Rebel politicians'] flight. The Northern press was frothing in print for their necks to be placed in the hangman's noose as traitors." The *New York Times*, for example, claimed the Rebels had committed "the greatest crime of the ages—a crime costing the lives of more than half a million men and aimed at the overthrow of the best government the world ever saw." With Lincoln's assassination, they could expect no mercy from East Tennessee's Andrew Johnson. He had previously described the Confederate hierarchy as an "illegitimate, swaggering, bastard, scrub aristocracy," and he had vociferously vowed to make them pay for the nation's four years of misery.[70]

On 2 April 1865, most of the Confederate Cabinet boarded "the last train from Richmond." The members included Pres. Jefferson Davis, of

---

[68] Ibid., 388–92.

[69] Grenier, *Central Florida's Civil War Veterans*, 8; M. E. Dickison, *Dickison and His Men*, 229–31; Murray, *Dixie's Land*, n.p.

[70] Hanna, *Flight into Oblivion*, xviii; David Cook, "Top Confederate leaders found refuge in Big Sun," *Ocala Star-Banner*, author collection, Rome, GA; Hollway, "Chasing Jefferson Davis."

Mississippi; Postmaster Gen. John H. Reagan, of Texas; Sec. of the Navy Stephen R. Mallory, of Florida; Attorney Gen. George Davis, of North Carolina; Sec. of the Treasury George A. Trenholm, of South Carolina; and, Sec. of State Judah P. Benjamin, of Louisiana. Sec. of War, John C. Breckinridge, of Kentucky, remained with Gen. Robert E. Lee's army until the last. (Confederate vice president Alexander H. Stephens of Georgia, always an outsider, had long since returned to his home state.) Two members of the cabinet—Breckinridge and Benjamin—would escape through Florida with the aid of the local partisans. George Davis also surrendered to the Federals at Key West, after enduring several harrowing weeks aboard a variety of boats, trying to reach the Bahamas.[71]

The Confederacy's fugitive chief executive waited a week at Danville, Virginia, on the North Carolina border, in a "last gasp" attempt to re-establish an outpost for the Confederate nation. Davis's small caravan then moved south. The war was lost, but Pres. Davis seemingly intended to turn west in an attempt to reach "Kirby Smith's Confederacy," in the Trans-Mississippi region. On the morning of 10 May, Federal cavalry captured the Davis party near Irwinville, Georgia. The Confederate president had donned a shawl against the morning chill, and the Northern press reported the bearded Mississippian had been wearing a dress. (As a result, a popular song sprang up in the wake of that spurious revelation, bearing the title "Jeff in Petticoats.") Davis would spend two years in prison, and an additional four years with a threat of a trial for treason hanging over his head.[72]

Judah P. Benjamin recognized, more clearly than any other Rebel cabinet member, the importance of escaping the pursuing Federals. In the wake of the assassination of Abraham Lincoln, "scoundrels and perjurers, attracted by the scent of reward money 'flocked like buzzards,' around the doors of the old Penitentiary...to swear that [Jefferson] Davis and Benjamin were the instigators of [John Wilkes] Booth and [Mrs. Mary] Surratt." The former Confederate secretary of war also recognized that he had to separate himself from the cumbersome collection of politicians, cavalry, with their wagon train, which was conspicuous in the hardscrabble, devastated countryside.[73]

---

[71] Hanna, *Flight into Oblivion*, 6–14; Hollway, "Chasing Jefferson Davis."

[72] Hollway, "Chasing Jefferson Davis"; B. Brown, "Capture of Jefferson Davis."

[73] Rosen, *Jewish Confederates*, 321–23; Cannon, "The Great Escape of Judah P. Benjamin."

Almost as soon as the fleeing Rebel politicians crossed the border into North Carolina, Benjamin took his leave from his cohorts. He had earlier created a false passport identifying himself as M. M. Bonfals. (The pseudonym clearly displayed Benjamin's quick mind and sardonic wit, as in Cajun French "bonfals" translates as "a good disguise.") To complete his transformation, he had grown a full beard, wore spectacles, and donned dirty, ragged clothes. Riding a nag, the second most important official in the Southern Confederacy meandered southward along Florida's west coast.[74]

Benjamin stayed briefly at his old friend Sen. David Yulee's Cottonwood Plantation, near Archer, and "for a short time in the home of his kinsman, Solomon Benjamin," near Ocala. Instinctively realizing that the Yankees would search the home of prominent Rebels, he stayed briefly at the humble abode of Capt. Leroy G. Leslie, near Brooksville. It seems difficult to imagine two more different allies—an urbane, likely agnostic Jewish attorney and politician, and a plain, "fire and brimstone" Methodist minister. Years later Maj. Munnerlyn attempted to explain their common ground. "Capt. Leslie" he wrote, "was quite an old man, but active and efficient, and a loyal Confederate." Their love for the fallen nation provided common ground for the two men. Leslie called together his friends, including James McKay Sr., and with Benjamin devised a plan to help the former secretary of war get to a foreign country. The southwest Florida Rebels suggested taking Benjamin to Cuba, but Benjamin objected. He feared "a weak Spanish Government might cave in to pressure from the powerful United State's [*sic*] demands for his return." Munnerlyn wrote, "After consultation it was concluded to obtain, if possible, a permit [from the Federals] to send a fishing smack to Nassau, under pretense of obtaining medical supplies." While they put their plan in motion, they moved Benjamin to more comfortable accommodations at the Gamble Mansion. It proved a dangerous miscalculation, and Benjamin only escaped capture by a Federal patrol by hiding in a the nearby patch of woods.[75]

Rev. Leslie and his co-conspirators set their scheme in motion immediately. They obtained a permit from the local Union commander, granting

---

[74] Rosen, *Jewish Confederates*, 321–22; Cannon, "The Great Escape of Judah P. Benjamin."

[75] "Judah P. Benjamin as Cook," *Charlotte Observer*, 10 June 1917, 13 (quoting a letter from Capt. Munnerlyn to his daughter regarding his war experiences); Rosen, *Jewish Confederates*, 322–23; Cannon, "The Great Escape of Judah P. Benjamin."

permission to make a "humanitarian" voyage, and the West Florida Rebels enlisted Capt. Frederick Tresca to pilot the vessel to Nassau. Tresca "had [often] run the blockade from Florida to Nassau and was thoroughly posted on the inside water route down the West Coast. He possessed cool courage, daring, and experience for such an adventure." Several harrowing events followed, including a near sinking and being battered by a terrific gale, but by 30 August 1865, Benjamin arrived in Great Britain. Though he had hardly a penny to his name, it did not take the former Confederate official long to become a highly respected barrister. When he died, nineteen years later, he was acknowledged as one of England's finest attorneys, author of the standard book on the sale of property, and an adviser to Queen Victoria.[76]

On 19 May, John C. Breckinridge crossed the Florida state line near Madison, accompanied by his aide, Col. James Wilson, and his body servant Tom Ferguson. (The former vice president of the Confederacy had earlier compelled his son to turn himself in to the Union troops, as mosquito bites caused a severe allergic reaction in the youth.) At Madison, Col. John Taylor Wood joined the Breckinridge party. Wood, the grandson of Pres. Zachary Taylor and an experienced mariner, would be a great help in the escape attempt. One thing Breckinridge did not share with anyone was how important it was for him to not fall into Federal hands. "General Sherman—at a meeting with Confederate officers following their capitulation,—advised Breckinridge to flee, explaining that the people in the Northern states, felt especially bitter toward the former U. S. Vice-President who took up arms against the government."[77]

Two days before the Federals would parole J. J. Dickison's troops, the Florida guerrilla leader received a note simply signed "Confederate Officer." Dickison found Breckinridge and his entourage at the home of Judge James B. Dawkins in Gainesville. The Kentuckian insisted he wanted to go to the Trans-Mississippi to join Gen. Kirby Smith. Dickison wisely suggested going to Cuba first, and then weigh his options. Dickison said he would have his men refloat the lifeboat from the *Columbine*, which had been secreted in a Central Florida lake. "I informed him," Dickison wrote, "of its [the boat's] capacity, twenty to twenty-five men, and told him if he would take the venture, this craft was at his service, and I would do all in my power to aid him in carrying out his heroic plan." The partisan leader also suggested moving

[76] Ibid.; Hanna, *Flight into Oblivion*, 225–27.

[77] Hanna, *Flight into Oblivion*, 42–43 and 105–106.

Breckinridge, as Alachua County would soon be swarming with Yankees coming to parole his men. During the night, some of Dickison's men moved the refugee to the plantation of Col. Samuel Owens, in Marion County.[78]

Breckinridge's party began their escape from the United States on the west bank of the St. Johns River, south of Palatka. They traveled up the waterway, constantly beset by swarms of gnats, mosquitoes, and a scorching sun that made the journey a miserable ordeal. At times they stayed in the middle of the larger lakes just for a little respite from the bloodsucking insects. When they reached Lake Harney, they decided to portage out of the swamps to the Indian River. The Rebel fugitives hired a mixed-blood local, with two cantankerous oxen, whose fumbling efforts to move the boat, brought a few moments of comic relief for the harried fugitives. John Taylor Wood recalled, "The next morning we made an early start. Our course was an easterly one through a roadless, flat sandy barren, with occasional thicket and swamp.... The waters of the Indian River were a most welcome sight and we hoped our troubles were over."[79]

The fugitives' trouble, however, had just begun. As they fled south, along the Atlantic coast, they encountered pirates, Union patrol boats, gales, and a lack of food and drinking water. With the *Columbine*'s lifeboat leaking badly, they commandeered at gunpoint a more seaworthy vessel, though they gave the "owners" a gold coin for "providing" the upgraded vessel. Breckinridge's faithful servant, Sam Ferguson, proved a godsend, and the Florida Confederates fondly remembered his cheerful outlook, quick mind in overcoming obstacles, and devotion to his master. Breckinridge, like Benjamin, wanted to reach Nassau, but Taylor wrote, "On the sixth day from the Florida coast we crossed the Nicolas Channel with fair wind. Soon we made the Cuban coast...[and] offered prayerful thanks for our wonderful escape."[80]

The fugitives landed at Cardenas, and some exiled Rebels and the locals made Breckinridge and his men comfortable. An early historian related, "The Confederates were allowed to pay for nothing." The group quickly moved on to Havana, and Breckinridge soon caught a ship to England. After remaining a few days on Cuban soil, Dickison's men returned to Central Florida. With

---

[78] M. E. Dickison, *Dickison and His Men,* 225; Klotter, *Breckinridges of Kentucky*, 129–31. See generally, Hanna, *Flight into Oblivion.*

[79] M. E. Dickison, *Dickison and His Men*, 225.

[80] Ibid., 225–27; Bell, *Confederate Seadog,* 115–38. The quote in this paragraph is from p. 119.

them they brought Sam Ferguson, the former slave whose devotion to Breckinridge had gained the admiration of the hard-bitten Rebels. Breckinridge traveled extensively, visiting France, the Holy Land, and England. The former Confederate secretary of war eventually settled in Canada until it proved safe to return to the Bluegrass State.[81]

[81] Bell, *Confederate Seadog*, 115–38; Hanna, *Flight into Oblivion*, 188–90.

Conclusion

"A Hideous and Hopeless Spectacle"

Florida did not join the new Rebel nation expecting to fight a guerrilla conflict. The Confederate government had pledged "[to maintain] state sovereignty, local defense, preservation of its territory, and maintenance of the geographical integrity of the South." The state, however, presented a number of unique problems for the new country. It had hundreds of miles of coastline, and much of the state, especially the southern region, was sparsely populated. Dr. David Coles noted "south Florida remained arguably the most remote area in the eastern United States."[1]

The Union high command initially dismissed the state as a placed of "little strategic importance." The Federals established military enclaves at Fernandina, St. Augustine, and a station for the "inner blockade" of the St. Johns River at Mayport. From those strongholds they made occasional raids into the countryside, but with no overall strategy for those forays. One of the earliest historians of Florida's conflict described "typical" military operations in the state.

> In East...South Florida the only plan followed by the Federal army was to harry and desolate the country whenever and wherever possible. Detachments of [Union] mounted troops moved often under cover of night and usually sought cotton, cattle, and personal effects [valuables]. The Confederate plan of resistance—if a plan it can be called—was to dog the course of the superior force with skirmishing from cover; or attempt by strategy and aggression to overwhelm smaller bodies [of Union raiders.]

The writer characterized that phrase of the war as "a hideous and hopeless spectacle."[2]

Despite early vows to protect every foot of the state, in spring 1862 the new nation's leaders bowed to military necessity and "abandoned" the state of Florida. A quick comparison of the Yankees and their Rebel opponents

---

[1] Sutherland, *Savage Conflict*, 53; Coles, "Cattle Wars," 95.

[2] Browning Jr., *Success Is All That Was Expected*, 16, 357; W. W. Davis, *Civil War and Reconstruction in Florida*, 297–98.

revealed why Floridians adopted guerrilla tactics to protect their families and property. The Federal forces—large, well equipped, and supported by infantry, cavalry, artillery, and gunboats—could have easily defeated any conventional army the Floridians could throw at them. A state, whose entire economy was based upon agriculture, could never hope to match the industrial might of the Federals, and irregular warfare became their tactic of last resort.[3]

Right after the Confederacy announced its decision to "abandon" Florida, Gov. John Milton declared, "One thousand men, divided into small companies [and] well-armed—acting as Guerillas or Rangers and ably commanded—can do more to defend Florida from the enemies than thousands in regular service." Milton recognized that men battling to protect their homes and families fight with greater motivation than an invading army. He also recognized that small units—even if they were poorly armed and supplied—could often defeat, or seriously damage, much larger forces. Acting upon that determination, Florida's chief executive began quietly organizing the Second Florida Cavalry (CSA), which soon formed the backbone Florida's defenses.[4]

Almost without exception the state's Rebel combatants went into each engagement outnumbered and outgunned, but they used ambush, rapid movement, and "hit and run" tactics to level the field. This method of combat required a "special kind of soldier." These individuals had certain characteristics that set them apart from more conventional fighting men. A modern scholar observed, "That they were all incredibly brave, and some of them foolhardy, can be taken for granted; but they also possessed those special traits of imagination, self-sufficiency, audacity, and determination. These qualities combined to produce a highly motivated soldier who performed well under independent command or as a small group under extraordinary and stressful conditions."[5]

Floridians seemed to have a natural affinity for irregular warfare. Dr. Sutherland noted, "One bright spot seemed to be the penchant of Florida [Rebels] for guerrilla warfare." Gov. Milton voiced a similar opinion, writing, "Many of our people have formed their idea of war...from experience derived

---

[3] Reiger, "Florida after Secession," 128–42; W. W. Davis, *Civil War and Reconstruction in Florida*, 297–98.

[4] John Milton, 23 April 1862, Milton Letterbook; Evans, ed. (and J. J. Dickison), *Confederate Military History* (Florida), 11:48–50.

[5] Pritchard, *Raiders of the Civil War*, 7.

from our Indian Wars." This allowed them to keep an eye open for "any improper conduction of the slave population, to prevent all communication between them and disloyal persons."[6]

During the first two years of the war, Florida's brush fighters did an admirable job of keeping the Federals confined in in their enclaves at Amelia Island and St. Augustine. The Rebel guerrillas faced a deeply divided population and the almost total lack logistical support from Tallahassee or Richmond, but their primary advantage was an intimate knowledge of the terrain. They knew shortcuts, fords, and hiding places it would take a lifetime to learn. Late in 1864, a Union soldier described the situation in a letter to a Northern newspaper:

> It is almost an impossibility for our forces to fight the rebels successfully the rebels successfully in Florida. In a country covered in extensive forests and immense swamps, having but few good roads, but thoroughly intersected by trails and bridle paths, whose whole surface is known to almost everyone of the enemy, but which our [Union} forces are totally unacquainted, it must continually occur that our forces will meet with loss and defeat while the enemy suffer comparatively little.[7]

A largely forgotten aspect of Florida's guerrilla actions involved its impact on the civilian population. A primary tenet of the shadow war states, "One essential factor of any successful [guerrilla] campaign…is a sympathetic supporting population." The success of Florida's partisans inspired the "home folks" to remain loyal to the Southern Confederacy, and "gave them reason to hope when most [war] news was bad." A modern writer noted, "Morale, both of the army and civilian population, is as important a part of the war efforts as generals and munitions, and it is in this area that the partisan rangers had their greatest impact." Leaders such as Dickison, Clarke, and others kept hope alive on the home front, even when news from Virginia and Tennessee often reported failure and defeat.[8]

---

[6] Sutherland, *Savage Conflict*, 114–15.

[7] "A Letter from the Department of the South, *Rutland [VT] Weekly Herald*, 8 September 1864, 2.

[8] Brownlee, *Gray Ghosts of the Confederacy*, 28; McLachlan, *American Civil War Guerrilla Tactics*, 59.

Florida's war changed in early 1863. The Emancipation Proclamation, issued following the Union victory at Antietam, allowed Union officers to begin enlisting Negroes into the Federal service. The Black men from Florida had earned a reputation for bravery and intelligence, and, as a result, raids up the St. Johns River and into the state's hinterlands increased. The Rebel guerrillas were often hard-pressed to halt a mass exodus of slaves, many of whom believed the biblical "Year of Jubilee" had arrived.[9]

With the fall of Vicksburg, Mississippi, and the Union conquest of several Southern "breadbasket regions," the Rebel armies turned to Florida to supply beef for the Confederate troops. Hundreds of thousands of half-wild longhorn cattle grazed in the South Florida, and a Rebel commissary officer for the Army of Tennessee addressed the situation bluntly, stating, "We are now dependent upon your state [Florida] for beef.... The future of the army depends on how well it is fed: and this in turn depends on our ability to secure food from Florida."[10]

It did not take the Federals long to recognize the importance of Florida beef. Northern journals published supposedly "confidential" letters regarding the Confederacy's dependence on Florida beef to feed its major armies, almost simultaneously with pro-Rebel journals. Union military leaders decided on a two-prong strategy to halt the delivery of beef to the Rebel armies. First, they made several attempts to establish a Union stronghold in East Florida to block the traditional routes of the cattle drovers. Both the battles of Olustee and the second fight at Gainesville were, at least partly, failed attempts to halt the flow of cattle from South Florida. A Union soldier from Vermont summed up the situation succinctly: "If the Federal troops could succeed only in cutting off the rebel supply of beef cattle from this state, it would be a most serious blow to this rebellion.... But we are sorry to record that thus far success has *not* crowned our efforts in these parts [emphasis added]."[11]

The second component of the Union plan to interdict the flow of South Florida was the enlistment of Negroes, Unionists, and Confederate deserters into the Federal army. They would be stationed in South Florida and conduct raids throughout the vast cattle region to disrupt the collection of herds at the

---

[9] See generally, Wilson, "In the Shadow of John Brown," 306. An excellent account of Black soldiers is Dobak's *Freedom by the Sword.*

[10] R. Taylor, "Rebel Beef," 15–20; Coles, "Far from Fields of Glory," 256–59.

[11] Coles, "Far from Fields of Glory," 256–59; "A Letter from the Department of the South."

source. As Dr. David Coles observed, "Not until late 1863 did the Federal military make an effort to maintain a permanent presence on the southern Florida mainland. When they did so the action came largely in response to an increased Confederate dependence on Florida cattle to feed the hungry armies of the South."[12]

Initially, that component of the plan worked well, with Union raids at Fort Meade and Brooksville panicking Confederate sympathizers in that region. The failure of the Fort Myers experiment resulted from the deeply held biases by White Southern Unionists and the United States Colored Troops. The disputes between those allies became so poisonous that Union commanders had no option but to separate the feuding units. Thereafter, the herding and driving cattle to the Rebel armies continued with few interruptions.[13]

At least part of the blame for the failure to halt the flow of Florida bovines must be leveled at the Federal high command in Washington. In late March 1864, Gen. Truman Seymour, who had commanded the Yankee army during the disastrous defeat at Olustee, requested to be relieved of command. "Seymour's relief led to a secession of commanders for North Florida, with none serving more than five consecutive months.... The frequent command changes undoubtedly affected Union morale, not to mention their impact on a coherent, defined strategy within the state." A couple of those officers, particularly Brig. Gen. William Birney, showed considerable initiative in leading raids into the state's interior; but none of them managed to establish a permanent bastion in the East Florida heartland.[14]

Meanwhile, the government in Richmond requested the state to "establish a military unit to defend Florida cattle herds from Union raids." This led to the formulation of the "Cow Cavalry," officially designated as the First Florida Special Cavalry Battalion, under the command of Maj. Charles J. Munnerlyn. Their primary job involved protecting the herds and driving cattle—and fighting only as a last resort. (These units contained a couple of exceptions, including Capt. Edward J. Lutterloh's Levy and Alachua Counties company, which operated primarily as a "home guard.")[15]

---

[12] Coles, "Cattle Wars," 95–96.

[13] C. Brown Jr., *Florida's Peace River Frontier*, 158–59; Dillon Jr., "Battle of Fort Myers," 28.

[14] Coles, "Far from Fields of Glory," 203–204.

[15] R. Taylor, "Cow Cavalry," 196–214.

As Gen. Robert E. Lee's Army of Northern Virginia surrendered at Appomattox Courthouse, cattle from Florida were nearing their destination at Petersburg. The state the Confederacy had abandoned in early 1862 continued to serve the Rebel nation to the bitter end.[16]

[16] R. Taylor, *Rebel Storehouse*, 159.

Bibliography

PRIMARY SOURCES

"An Act to Exempt certain persons from military service…" (approved 11 October 1862). *Public Laws of the Confederate States of America Passed at the First Session of the First Congress.* Richmond, VA: James M. Mathews Publishers, 1862.

Bethell, John A. "Pinellas: A Brief History of the Lower Point." 1914. City, County, and Regional Histories E-Book Collection. 1. https://digitalcommons.usf.edu/regional_ebooks/1.

Blakey, Arch Frederic, et al. *Rose Cottage Chronicles: Civil War Letters of the Bryant-Stephens Families of North Florida.* Gainesville: University of Florida Press, 1998.

Boggess, Francis C. M. *A Veteran of Four Wars: The Autobiography of F. C. M. Boggess.* Arcadia, FL: Champion Job Room, 1900.

Burtchaell, W. D. "Capture of a Vessel Off of Cedar Keys." Unpublished manuscript. Confederate Miscellany, 18–20, Georgia Archives and History, Morrow, GA.

———. "Recapture of Four Vessels Off Cedar Keys." Unpublished manuscript. Confederate Miscellany, 13–17, Georgia Archives and History, Morrow, GA.

Chesnut, Mary Boykin. *A Diary from Dixie.* New York: Appleton, 1905.

Croffut, W. A., and John M. Morris. *The Military History of Connecticut during the War of 1861–1865.* New York: Ledyard Bill, 1868.

Evans, Clement A., ed. *Confederate Military History.* Vol. 11: Texas and Florida; Florida section by J. J. Dickison. Secaucus, NJ: The Blue & Grey Press, 1976.

Denison, Frederic. *Shot and Shell: The Third Rhode Island Heavy Artillery Regiment in the Rebellion, 1861–1865.* Providence: Third R.I.H. Art. Vet. Association, by J.A. & R.A. Reid, 1879.

Dickison, Mary Elizabeth. *Dickison and His Men: Reminiscences of the War in Florida* (facsimile). Jacksonville, FL: San Marco, 1962.

Du Pont, Samuel Frances. *Samuel Frances Du Pont: A Selection of His Civil War Letters.* 3 vols. Edited by John D. Hayes. Ithaca, NY: Cornell University Press, 1969.

Ellis, Thomas Benton. "Diary, July 1861–April 1865." Gainesville: P.K. Younge Library of Florida History, University of Florida.

Fort, John Porter. *A Memorial and Personal Reminiscences.* New York: Knickerbocker Press, 1918.

Higginson, Thomas Wentworth. *Army Life in a Black Regiment.* East Lansing: Michigan State University, 1960.

———. *The Complete War Journal and Selected Letters of Thomas Wentworth Higginson.* Edited by Christopher Looby. Chicago: University of Chicago Press, 2000.

Holn, John. *Tison Exposa*. Pamphlet.

Hope, Capt. Samuel E. "A Report on S. E. Hope's Part of a Civil War Incident Near Jacksonville." Typescript. Private collection of Zack C. Waters, Rome, GA.

Lesley, Theodore. "The Organization of the Confederate Cattle Battalion of Florida." Theodore Lesley Papers, collection of Vernon Peeples, Punta Gorda, FL.

McCullough, William. "My National Troubles: The Civil War Papers of William McCullough." Edited by Kyle VanLandingham. *Sunland Tribune* 20 (1994): 59–86.

McMurray, Richard M., ed. *Footprints of a Regiment: A Recollection of the 1st Georgia Regulars*. Atlanta: Longstreet Press, 1992.

Milton, Governor John. Letterbook, 1861–1865. State Archives of Florida, Florida Memory. 13 March 1862; 24 March 1862; 3 April 1862; 23 April 1862; 25 April 1862. Cited as Milton Letterbook.

"The Oklawaha Rangers." unpublished mss. [by Pearson's daughter]. Zack C. Waters collection, Rome, GA.

Official Records of the Union and Confederate Navies in the War of Rebellion. 30 vols. Washington, DC: Government Printing Office, 1894–1922. Cited as *ORN*.

Patterson, Edmund DeWitt. *Yankee Rebel: The Civil War Journal of Edmund DeWitt Patterson*. Chapel Hill: University of North Carolina Press, 1966.

"The Oklawaha Rangers." Unpublished manuscript. In the private collection of Zack C. Waters, Rome, GA.

Public Laws of the Confederate States of America. Richmond, VA: John M. Matthews Publishers, 1862.

Reid, Whitelaw. *Ohio in the War: Her Statesmen, Generals, and Soldiers*. Vol. 2. Cincinnati: Robert Clarke Co., 1895.

Rerick, Roland H. *Memoirs of Florida*. Vol. 1. Atlanta, GA: Southern Historical Society, 1902.

Robinson, Calvin L. "An Account of Some of My Experiences in Florida during the Rise and Progress of the Rebellion." Typescript. Jacksonville Historical Society, Jacksonville, FL.

Roe, Alfred S. *The Twenty-fourth Regiment of Massachusetts Volunteers*. Worchester, MA: Twenty-fourth Veterans, 1907.

Russell, William Howard. *My Diary, North and South*. London: Bradbury and Evans, 1863.

Shaw, Robert G., to Annie Kneeland Haggerty Shaw, 12 June 1863.

Southerton, William B. "Reminisces of the Civil War." Manuscript. Columbus, OH: Ohio Historical Society, 1937.

War of the Rebellion: A Compilation of the Official Records of the Union and Confederate Armies. 70 vols. In 128 parts. Washington, D.C.: Government Printing Office, 1880–1901. Cited as *OR*.

SECONDARY SOURCES

Akerman, Joe A. *Florida Cowman: A History of Florida Cattle Raising*. Kissimmee, FL: Florida Cattlemen's Association, 1976.

———, and J. Mark Akerman. *Jacob Summerlin: King of the Crackers*. Cocoa, FL: Florida Historical Society, 2004.

Allardice, Bruce S. *More Generals in Gray*. Baton Rouge: Louisiana State University Press, 1995.

Asarch, Rhonda V. "Liberty Billings: Florida's Forgotten Radical Republican." Unpublished master's thesis, Florida Atlantic University, 2012.

Ash, Stephen V. *When the Yankees Came: Conflict and Chaos in the Occupied South*. Baton Rouge: Louisiana State University Press, 1988.

Baggett, James Alex. *The Scalawags: Southern Dissenters in the Civil War and Reconstruction*. Baton Rouge: Louisiana State University Press, 2004.

Baltzell, Col. George F. "The Battle of Olustee." *Florida Historical Quarterly* 9 (1931): 199–223.

Bearss, Edwin C. "Military Operations on the St. Johns, September–October, 1862." *Florida Historical Quarterly* 42 (1964) pt. 3, 233–48; 42 (1964) pt. 4, 332–51.

Bell, John. *Confederate Seadog: John Taylor Wood in War and Exile*. Jefferson, NC: McFarland, 2002.

Berlin, Ira, et al. *The Destruction of Slavery*. Ser. 1, vol. 1 of Freedom: A Documentary History of Emancipation, 1861–1867. New York: Cambridge University Press, 1985.

Blakey, Arch Frederic, and Bonita Thomas Deaton. *Parade of Memories: A History of Clay County, Florida*. Green Cove Springs, FL: Clay County Bicentennial Steering Committee, 1976.

Boot, Max. *Invisible Armies: An Epic History of Guerrilla Warfare from Ancient Times to the Present*. New York: W. W. Norton & Co., 2013.

Bradshaw, Jim. "Armies of Last Resort." *America's Civil War* 25 (2012): 30–35.

Broadwater, Robert P. *The Battle of Olustee, 1864: The Final Union Attempt to Seize Florida*. Jefferson, NC: McFarland, 2006.

Broward, Robert C. *The Broward Family: From France to Florida, 1764–2011*. Jacksonville, FL: Jacksonville Historical Society, 2011.

Brown, Canter, Jr. *Florida's Peace River Frontier*. Orlando: University of Central Florida Press, 1991.

———. *In the Midst of All That Makes Life Worth Living: Polk County, Florida to 1940*. Tallahassee: Sentry Press, 2001.

———. "Tampa's James McKay and the Frustration of Confederate Cattle Supply Operations in South Florida." *Florida Historical Quarterly* 70/4 (April 1992): 409–33.

Browning, Robert M. *Success Is All That Was Expected: The South Atlantic Blockading Squadron during the Civil War*. Washington, DC: Potomac Books, 2003.

Brownlee, Richard S. *Gray Ghosts of the Confederacy: Guerrilla Warfare in the West, 1861–1865.* Baton Rouge: Louisiana State University Press, 1983.

Buker, George E. *Blockaders, Refugees, & Contrabands: Civil War on Florida's Gulf Coast, 1861–1865.* Tuscaloosa: University of Alabama Press, 1993.

———. "The Inner Blockade of Florida and the Wildcat Blockade Runners." *North & South* 4/2 (2001): 70–85.

Burdett, Susan. "The Military Career of Brigadier General Joseph Finegan, CSA, of Florida." Master's thesis, Columbia University, New York, 1930.

Caldwell, Harold D., Sr. "New Smyrna: Confederacy's Keyhole Through the Union Blockade." In *The Civil War in Volusia County: A Symposium.* Daytona: Halifax Historical Society, 1987.

Cash, W. T. "Taylor County History and Civil War Deserters." *Florida Historical Quarterly* 27/1 (July 1948): 28–58.

Coles, David J. "Ancient City Defenders: The St. Augustine Blues." In *Civil War Times in St. Augustine,* edited by Jacqueline K. Fretwell, 65–89. Port Salerno: Florida Classics Library, 1988.

———. "Cattle Wars in the Civil War in South Florida." *The Proceedings of the Florida Cattle Frontier Symposium.* Kissimmee, FL: Florida Cattlemen's Association and Florida Cracker Cattle Association, 2003.

———. "Far from Fields of Glory: Military Operations in Florida during the Civil War, 1864–1865." PhD dissertation, Florida State University, 1996.

———. *Florida Civil War Heritage Trail.* Tallahassee: Florida Association of Museums, 2011.

———, and Zack C. Waters. "Indian Fighter, Confederate Soldier, Blockade Runner, and Scout: The Life and Letters of Jacob E. Mickler." *El Escribano* 34 (1997): 35–69.

———. "Volusia County: The Land Warfare, 1861–1865." *The Civil War in Volusia County: A Symposium.* Daytona Beach, FL: Halifax Historical Society, 1987.

Collins, Traci C. "The Saga of the Steamboat *Madison.*" *Hidden Coastlines* (Summer 2013): 16–18.

Cooper, William J. *We Have War upon Us: The Onset of the Civil War, November 1860–April 1861.* New York: Alfred A. Knopf, 2012.

Covington, James W. *The Story of Southwestern Florida.* New York: Lewis Historical Publishing, 1957.

Davis, William C., ed. *The Confederate General.* 6 vols. Washington, DC: National Historical Society, 1991.

Davis, William W. *The Civil War and Reconstruction in Florida.* Gainesville: University of Florida Press, 1964.

De La Cova, Antonio Rafael. *Colonel Henry Theodore Titus: Ante-Bellum Soldier of Fortune and Florida Pioneer.* Columbia: University of South Carolina Press, 2016.

Dillon, Rodney E., Jr. "'A Gang of Pirates': Confederate Lighthouse Raids in Southeast Florida, 1861." *Florida Historical Quarterly* 67/4 (April 1989): 441–57.

———. "The Battle of Fort Myers." *Tampa Bay History* 5 (1983): 27–36.

———. "The Civil War in South Florida." Unpublished master's thesis, University of Florida, 1980.

———. "The Little Affair: The Southwest Florida Campaign, 1863–1864." *Florida Historical Quarterly* 63 (1984): 316.

Dobak, William A. *Freedom by the Sword: The U. S. Colored Troops, 1862–1867.* New York: Skyhorse Publishing, 2013.

DuBois, Bessie Wilson. "Two Florida Lighthouse Keepers." *Tequesta* 33 (1973): 41–50.

Dye, R. Thomas. "Defeated in War and Peace: The Political and Military Career of Major Edmund C. Weeks." *Sunland Tribune* 22 (1996): 45–54.

East, Omega G. "St. Augustine during the Civil War." *Florida Historical Quarterly* 31/2 (1952): 75–91.

Fishburne, Charles Carroll, Jr. *The Cedar Keys in the 19th Century.* Cedar Key, FL: Cedar Key Historical Society, 2004.

Fitzgerald, T. E. *Volusia County: Past and Present.* Daytona Beach, FL: Observer Press, 1937.

Futch, Ovid Leon. "Salmon P. Chase and Civil War Politics in Florida." *Florida Historical Quarterly* 32/3 (January 1952): 163–88.

Glatthaar, Joseph T. *Forged in Battle: The Civil War Alliance of Black Soldiers and White Officers.* Baton Rouge: Louisiana State University Press, 1990.

———. *General Lee's Army: From Victory to Collapse.* New York: Free Press, 2008.

Gleeson, Ed. *Erin Go Gray!: An Irish Rebel Trilogy.* Carmel, IN: Guild Press of Indiana, 1997.

Gould, Pleasant D. *History of Volusia County, Florida.* Deland, FL: Painter, 1927.

Graff, Mary B. *Mandarin on the St. Johns.* Gainesville: University of Florida Press, 1953.

Graham, Thomas. "The Home Front: Civil War Times in St. Augustine." In *Civil War Times in St. Augustine*, edited by Jacqueline K. Fretwell, 19–45. Port Salerno, FL: Florida Classics Library, 1988.

Grenier, Bob. *Central Florida's Civil War Veterans.* Charleston, SC: Arcadia Publishing, 2014.

Griffin, Patricia C. *Mullet on the Beach: The Minorcans in Florida, 1768–1788.* Gainesville: University of Florida Press, 1991.

Grismer, Karl H. *A History of the City of Tampa and the Tampa Bay Region of Florida* [McKay, D. B., ed.]. St. Petersburg: St. Petersburg Printing, 1950.

Grzelak, Jeff H. *Hernando County during the Civil War.* Orlando, FL: UADF Publications, 1987.

Hannah, Alfred Jackson. *Flight into Oblivion.* Richmond, VA: Johnson Publishing, 1938.

Hannings, Bud. *Every Day of the Civil War: A Chronological Encyclopedia.* Jefferson, NC: McFarland, 2010.

Harris, William Lloyd. "When War Came to Polk County: Battle at Bowlegs Creek—February 20, 1864." *The United Daughters of the Confederacy Magazine* 61 (1988): 28–29.

Hartman, David W., and David J. Coles. *Biographical Roster of Florida's Confederate and Union Soldiers.* 6 vols. Wilmington, NC: Broadfoot, 1995. Cited as *BRF.*

Hawk, Robert. *Florida's Army: Militia/State Troops/National Guard, 1565–1985.* Englewood, FL: Pineapple Press, 1986.

Hawley, Jennifer J. "Florida's Civil War Soldiers." Unpublished master's thesis, University of South Florida, 2005.

Hebel, Ianthe Bond. *Centennial History of Volusia County, Florida, 1854–1954.* Daytona Beach, FL: College Publishing Co., 1955.

Hemming, Charles C. "A Confederate Odyssey." *American Heritage* 36/1 (1984): 69–84.

Hewitt, Lawrence Lee. "Forgotten Captive." *America's Civil War* 29 (2016): 44–51.

———. "William Miller." In *Confederate General,* edited by W. C. Davis, 4:176–177. Washington, DC: National Historical Society, 1991.

Hillhouse, Don. *Heavy Artillery & Light Infantry: A History of the 1st Florida Special Battalion & 10th Infantry Regiment, CSA.* Jacksonville, FL: n. p., 1992.

Horn, Stanley F. *The Army of Tennessee.* Wilmington, NC: Broadfoot, 1987.

Ivey, Donald J. "The Accidental Pioneer: Capt. James McMullen and the Taming of the Pinellas Peninsula." *Sunland Tribune* 22 (1996): 27–40.

———. "John T. Lesley: Tampa's Pioneer Renaissance Man." *Sunland Tribune* 21 (1995): 3–20.

Joes, Anthony James. *America and Guerrilla Warfare.* Lexington: University of Kentucky Press, 2004.

Johannes, Jan H. *Yesterday's Reflections II: Nassau County, Florida.* Fernandina Beach, FL: Lexington Ventures, 2000.

Jones, James P., and William Warren Rogers. "The Surrender of Tallahassee." *Apalachee* 6 (1963–1967): 103–110.

Jones, Terry L. "James Heyward Trapier." In *Confederate General,* ed. W. C. Davis, 6:58–59. Washington, DC: National Historical Society, 1991.

King, Jerry McDowell. "Shadow Warriors of the Confederacy." Unpublished master's thesis, University of South Florida, 1990.

Klotter, James C. *The Breckinridges of Kentucky, 1760–1981.* Lexington: University Press of Kentucky, 1986.

Knetsch, Joe, and Paul S. George. "A Problematical Law: The Armed Occupation Act of 1842 and its Impact on Southeast Florida." *Tequesta* 53 (1993): 63–80.

———. "Armed Boats off Crystal River: A Chapter of the Civil War in Citrus County." *At Home: Citrus County Historical Society* 18 (2001): 6–7.

———. "Forging the Florida Frontier: The Life and Career of Captain Samuel Hope." *Sunland Tribune* 20 (1994): 31–40.

———. "John Westcott and the Coming of the Third Seminole War: Perspectives from Within." Paper presented to the Florida Historical Society Annual Meeting, 12 May 1990. Personal collection of Zack C. Waters, Rome, GA.

———. "Madison Starke Perry and the Spirit of Secession." Paper delivered to the Florida division of the Sons of Confederate Veterans, 20 June 1988, Lake City, Florida. Personal collection of Zack C. Waters, Rome, GA.

Koblas, John J. *John J. Dickison: Swamp Fox of the Confederacy.* St. Cloud, MN: North Star Press, 2000.

Krick, Robert K. *Lee's Colonels: A Biographical Register of Field Officers in the Army of Northern Virginia.* Dayton, OH: Morningside, 1991.

Litrico, Charles. "Joseph Finegan: Fernandina's Confederate General." *Amelia Now* (Fall, 2011), 23–27.

Loderhose, Gary. *Way down upon the Suwannee River: Sketches of Florida during the Civil War.* Lincoln, NE: Author's Choice Press, 2000.

Martin, Richard A., and Daniel Schafer. *Jacksonville's Ordeal by Fire: A Civil War History.* Jacksonville: Florida Publishing Co., 1984.

McCarthy, Kevin. *Cedar Key, Florida: A History.* Charleston, SC: History Press, 2007.

———. *St. Johns River Guidebook.* Sarasota, FL: Pineapple Press, 2004.

McLachlan, Sean. *American Civil War Guerrilla Tactics.* New York: Osprey, 2009.

McMurray, Richard M. "The President's Tenth and the Battle of Olustee." *Civil War Times Illustrated* 16 (1978): 12–25.

Michaels, Brian E. *The River Runs North: A History of Putnam County, Florida.* Palatka, FL: Taylor Publishing Co., 1986.

Michaels, Will. *The Making of St. Petersburg.* Charleston, SC: History Press, 2012.

Mosocco, Ronald A. *The Chronological Tracking of the American Civil War per Official Records of the War of the Rebellion.* Williamsburg, VA: James River Publications, 1993.

Murfree, Boyd R. "Rebel Sovereigns: The Civil War Leadership of John Milton of Florida and Joseph E. Brown of Georgia, 1861–1865." PhD dissertation, Florida State University, 2009.

Murray, Vince. *Dixie's Land.* Ocala, FL: Florida Focus Publications, 1983.

Nulty, William H. *Confederate Florida: The Road to Olustee.* Tuscaloosa: University of Alabama Press, 1990.

Ott, Eloise Robinson, and Louis Hickman Chazal. *Ocali Country: Kingdom of the Sun: A History of Marion County, Florida.* Ocala, FL: Marion Publishers, 1974.

Page, Dave. *Ship versus Shore: Civil War Engagements along the Shores and Rivers.* Nashville, TN: Rutledge Hill Press, 1994.

Pankenier, Charles. *Ridgeland Fights the Civil War.* New York: Worthy Shorts, 2011.

Pizzo, Anthony P. *Tampa Town: 1824–1886; The Cracker Village with a Latin Accent.* Miami: Hurricane House, 1986.

Porter, Tania Mosier, and Cassandra Fyotek. *The Story of Orlando and Orange County.* San Antonio: Historical Publishing Network, 2009.

Price, Marcus W. "Ships That Tested the Blockade of Georgia and East Florida Ports, 1861–1865." *American Neptune* 15 (1955): 97–132.

Pritchard, Russ A. *Raiders of the Civil War: Untold Stories of Actions Behind the Lines.* Guilford, CT: Lyon Press, 2005.

Proctor, Samuel, ed. *Florida: A Hundred Years Ago.* Tallahassee: Florida State Library, 1961–1965.

———. "Jacksonville during the Civil War." *Florida Historical Quarterly* 41/4 (April 1963): 344–56.

Ramage, James A. *Rebel Raiders: The Life of General John Hunt Morgan.* Lexington: University of Kentucky Press, 1996.

Redd, Robert. *St. Augustine in the Civil War.* Charleston, SC: History Press, 2014.

Reiger, John F. "Deprivation, Disaffection, and Desertion in Confederate Florida." *Florida Historical Quarterly* 48/3 (January 1970): 279–98.

———. "Florida after Secession: Abandonment by the Confederacy and its Consequences." *Florida Historical Quarterly* 50/2 (October 1971): 129–43.

Robertson, Fred L., comp. *Soldiers of Florida in the Seminole Indian, Civil, and Spanish-American Wars.* 1903. Mcclenny, FL: Richard J. Ferry, 1983.

Rosen, Robert N. *The Jewish Confederates.* Columbia: University of South Carolina Press, 2000.

Schafer, Daniel L. "Freedom Was as Close as the River." In *African American Heritage of Florida*, edited by David R. Colburn and Jane L. Landers, 157–84. Gainesville: University Press of Florida, 2018.

———. *Thunder on the River: The Civil War in Northeast Florida.* Gainesville: University of Florida Press, 2010.

Schmidt, Lewis G. *Florida's East Coast.* The Civil War in Florida: A Military History. Vol. 1, part 1. Allentown, PA: Schmidt, 1991.

———. *The Battle of Olustee.* The Civil War in Florida: A Military History. Vol. 2. Allentown, PA: Schmidt, 1989.

Levy County Archives Committee. "The Yearty Family." In vol. 11 [also called ch. 11] of *Search for Yesterday: A History of Levy County, Florida.* Bronson, FL: Levy County Board of Commissioners, 1982.

Sheppard, Jonathan C. *By the Noble Daring of Her Sons: The Florida Brigade of the Army of Tennessee.* Tuscaloosa: University of Alabama Press, 2012.

Shingleton, Royce. *High Seas Confederate: The Life and Times of John Newland Maffitt.* Columbia: University of South Carolina Press, 1954.

Sifakis, Stewart. *Compendium of the Confederate Armies: Florida and Arkansas.* New York: FactsOnFile, 1992.

Stephens, Arnold Dix. *Withlacoochee Notes.* Trenton, FL: n. p., 2005.

Still, William N. "A Naval Sieve: The Union Blockade in the Civil War." *Naval War College Review* 36 (1983): 38–45.

Stoutmire, William F. "Florida's Army: Home Guard Cavalry and Confederate Supply, 1861–1865." Master's thesis, Florida State University, 2008.

Strickland, Alice. "Blockade Runners." *Florida Historical Quarterly* 36/2 (October 1957): 86–94.

Sutherland, Daniel E. *A Savage Conflict: The Decisive Role of Guerrillas in the Civil War.* Chapel Hill: University of North Carolina Press, 2009.

Sweett, Zelia Wilson. *New Smyrna, Florida in the Civil War.* New Smyrna, FL: Volusia County Historical Commission, 1963.

Sword, Wiley. *Firepower from Abroad: The Confederate Enfield and the LeMat Revolver.* Man at Arms Monograph Series. Lincoln, RI: Mowbray, 1986.

———. *The Confederacy's Last Hurrah: Spring Hill, Franklin, and Nashville.* Lawrence: University of Kansas Press, 1992.

Taylor, Paul. *Discovering the Civil War in Florida: A Reader's Guide.* Sarasota, FL: Pineapple Press, 2001.

Taylor, Robert A. "Cow Cavalry: Munnerlyn's Battalion in Florida, 1864–1865." *Florida Historical Quarterly* 65/2 (October 1986): 196–214.

———. "Rebel Beef: Florida Cattle and the Confederate Army, 1862–1864." *Florida Historical Quarterly* 67/1 (1988): 15–31.

———. *Rebel Storehouse: Florida in the Confederate Economy.* Tuscaloosa: University of Alabama Press, 1995.

Totten, Eric Paul. "The Ancient City Occupied: St. Augustine as a Test Case for Stephen Ash's Civil War Occupied Model." Master's thesis, University of Central Florida, 2011.

VanLandingham, Kyle S. "The Union Occupation of Tampa, May 6–7, 1864." *Sunland Tribune* 19 (1993): 9–16.

———. "William B. Hooker: Florida Cattle King." *Sunland Tribune* 22 (1996): 3–18.

Warner, Ezra G. *Generals in Blue: Lives of Union Commanders.* Baton Rouge: Louisiana State University Press, 1992.

Waters, Zack C. "The Enigmatic Colonel Harry Maury and the Fifteenth Confederate Cavalry." *Alabama Heritage* 90 (Fall 2010) 8–17.

———. "Florida's Confederate Guerrillas: John W. Pearson and the Oklawaha Rangers." *Florida Historical Quarterly* 70/2 (October 1991): 133–49.

———. "Our Necessary Resort: Confederate Guerrillas in East and Middle Florida." In *A Forgotten Front: Florida in the Civil War Era*, edited by Seth A. Weitz and Jonathan C. Sheppard, 72–88. Tuscaloosa: University of Alabama Press, 2018.

———, and James C. Edmonds. *A Small but Spartan Band: The Florida Brigade in Lee's Army of Northern Virginia.* Tuscaloosa: University of Alabama Press, 2010.

———. "Tampa's Forgotten Defenders: The Confederate Commanders of Fort Brooke." *Sunland Tribune* 17 (1991): 3–12.

———. "'Tell Them I Died Like a Confederate Soldier': Finegan's Florida Brigade at Cold Harbor." *Florida Historical Quarterly* 69/2 (October 1990): 156–77.

Weinert, Richard P. "Dickison—The Swamp Fox of Florida." *Civil War Times Illustrated* 5 (1966): 4–11.

Wilson, Keith. "In the Shadow of John Brown: The Military Service of Colonels Thomas Higginson, James Montgomery, and Robert Shaw in the Department of the South." In *Black Soldiers in Blue: African American Troops in the Civil War*, edited by John David Smith, 306–35. Chapel Hill: University of North Carolina Press, 2002.

Winsboro, Irvin D. S. *Florida's Civil War: Explorations into Conflict, Interpretation, and Memory.* Cocoa: Florida Historical Society, 2007.

Wise, Stephen R. *Lifeline of the Confederacy: Blockade Running during the Civil War.* Columbia: University of South Carolina Press, 1988.

Wynne, Lewis N., and Robert Taylor. *Florida in the Civil War*. Charleston, SC: Arcadia Publishing, 2002.

NEWSPAPERS (BY STATE)

California
Sacramento Daily Union
San Francisco Bulletin
Connecticut
Hartford Courant
Hartford Daily Courant
New Haven Palladium
Stamford Advocate
District of Columbia
Washington Daily National Intelligencer
Washington Daily National Republican
Washington Evening Star
Florida
Gainesville Cotton Plant
Ocala Banner
Ocala Star-Banner
Palatka News and Advertiser
St. Augustine Evening Record
Georgia
Atlanta Southern Confederacy
Augusta Daily Chronicle
Augusta Daily Constitutionalist
[Augusta] Southern Christian Advocate

Columbus Daily Enquirer
Macon Telegraph
Illinois
Springfield Illinois State Journal
Indiana
Evansville Daily Journal
Iowa
Council Bluffs Weekly Nonpareil
Kentucky
Lexington Leader
Louisiana
New Orleans Daily True Delta
New Orleans Times-Picayune
Maine
Portland Daily Eastern Argus
Maryland
Baltimore Sun
Elkton Cecil Whig
Massachusetts
Boston Evening Transcript
Boston Herald
Boston Post
Boston Traveler
Lowell Daily Citizen & News
Springfield Republican
Worchester Aegis and Transcript
Worchester Massachusetts Spy
Worchester National Aegis
Michigan
Jackson Citizen
Kalamazoo Gazette
New Hampshire
Amherst Farmer's Cabinet
Manchester Daily Mirror
Manchester Weekly Union
New York
Albany Evening Journal
Jamestown Journal
New York Commercial Advertiser
New York Daily People
New York Evening Post
(New York City) Frank Leslie's Illustrated Newspaper

New York Herald
New York Times
New York Tribune
New York World
New York Weekly Anglo-African
North Carolina
Fayetteville Carolina Observer
Fayetteville Observer
Ohio
Cleveland Plain Dealer
Pomeroy Weekly Telegraph
Pennsylvania
Butler American Citizen
Harrisburg Patriot
Philadelphia Dollar Newspaper
Philadelphia Inquirer
Philadelphia North American
Philadelphia Press
Philadelphia Public Ledger
Washington Reporter
Rhode Island
Providence Evening Press
South Carolina
Charleston Courier
Charleston Mercury
Lancaster Ledger
Port Royal New South
Port Royal Palmetto Herald
Tennessee
Memphis Daily Appeal (published at various Southern cities)
Nashville Daily Union
Vermont
Danville North Star
Montpelier Daily Green Mountain Freeman
Rutland Weekly Herald
Virginia
Richmond Enquirer
Richmond Examiner
Richmond Sentinel
Richmond Whig
Wisconsin
Jackson Citizen

Milwaukee Sentinel
New Lisbon Juneau County Argus

WEBSITES
"1843. Ernestine (Strong) Alberti to Lydia (Strong) Clapp." William Griffing, transcriber. Blog. https:sparedandshared.wordpress.com/letters/1843-ernestine-strong-alberti-to-lydia-strong-clapp/. Accessed 27 June 2022.
Adams, Lillian. "The Baldwin Story: Baldwin, Florida." 1976. http://www.cow-art.info/documents/Baldwin_History0001.pdf. Accessed 24 June 2022.
Bamford, David M., and Kyle S. VanLandingham. "Now Just Hold Your Cow Cavalry Horses! A Rebuttal of Irvin D. Solomon and Grace Erhart's Vicious Smear of Munnerlyn's Cattle Guard Battalion." http://www.oocities.org/yes_album/Rebuttal.html. Accessed 27 June 2022.
Barker-Benfield, Simon. "Aug. 9, 1864: Union regiments' raid from Baldwin proved less than successful." 9 August 2019. https://www.jacksonville.com/story/special/special-sections/2019/08/09/aug-9-1864-union-regiments-raid-from-baldwin-proved-less-than-successful/4494203007/.
Beard, Rick. "$10 a Month: Men of the USCT." 24 February 2017. https://www.historynet.com/10-month-men-USCT/?f.
Bowden, Denny. "Volusia's Bloodiest Civil War Battles." Volusia History: Tracing Florida's Past. 12 January 2014. https://volusiahistory.wordpress.com/2014/01/12/volusias-bloodiest.civil-war-battle/.
"Brigadier General Truman Seymour, U.S.A. (1824–1891)." Olustee Battlefield Citizens Support Organization, Inc. https://battleofolustee.org/seymour.html. Accessed 24 June 2022.
Brown, Brian. "Capture of Jefferson Davis." New Georgia Encyclopedia. Updated 6 June 2017. https://www.georgiaencyclopedia.org/articles/history-archaeology/capture-of-jefferson-davis/.
"Buffington Family." https://www.drbronsontours.com/bronsonbuffingtonfamily.html.
"Calendar of Civil War Activity 1862–1864." Clay County Historical Archives, Clay County, Florida. http://archives.clayclerk.com/county-history/time-period/civil-war-reconstruction/. Accessed 24 June 2022.
Cannon, Jeff. "The Great Escape of Judah P. Benjamin." Updated 17 January 2012. https://patch.com/florida/newportrichey/the-great-escape-judah-p-benjamin.
"Capt. James Dopson Green, Company B, Second Florida Cavalry, USA." https://freepages.rootsweb.com/~crackerbarrel/genealogy/Green%20Letters.html. Accessed 24 June 2022.
"The Civil War Home Page." www.civil-war.net/pages/1860_census.html.
"Col. Louis Bell." www.drbronsontours.com/bronsoncolloubellbio.html.

"The Confederate Swamp Fox." Museum of Southern History. Jacksonville, FL. https://www.museumsouthernhistory.com/The%20Confederate%20Swamp%20Fox.html. Accessed 27 June 2022.

Cox, Dale. "Apalachicola Arsenal, Chattahoochee, Florida." Updated 4 January 2015. http://www.exploresouthernhistory.com/arsenal1.html.

———. "Battle of Gainesville; Gainesville, Florida; Florida's Swamp Fox does battle!" Updated 24 May 2017. http://www.exploresouthernhistory.com/gainesvillebattle.html.

———. "The Civil War in Florida: Skirmish at New Smyrna, March 22, 1862."

———. "Gamble Plantation Historic State Park—Ellenton, Florida." 2011. www.exploresouthernhistory.com/gamble.html.

———. "The Union Attack on Cedar Key—Cedar Key, Florida." 2012. https://www.exploresouthernhistory.com/cedarkeyattack.html. Accessed 27 June 2022.

"Crystal River National Wildlife Refuge." US Fish & Wildlife Service. https://www.fws.gov/crystalriver. Accessed 24 June 2022.

Edling, Richard. "Fernandina Beach, Florida." http://civilwaralbum.com/misc5/fernandina_beach1.htm. Accessed 24 June 2022.

Ekardt, David. "The Civil War: The Navy's Great Salt Raids." The Civil War. August 2009. http://www.navyandmarine.org/ondeck/1862saltraids.htm.

Erlandson, Marcus R. "Guy V. Henry: A Study in Military Leadership." Master's thesis, University of Wisconsin, 1980. https://apps.dtic.mil/sti/pdfs/ADA164528.pdf.

Filzen, Lydia Colee. "Brooksville Raid." Museum of Southern History. Jacksonville, FL. https://www.museumsouthernhistory.com/Brooksville%20Raid.html. Accessed 24 June 2022.

"Florida 2nd Cavalry, Co. K." Information provided by Cosette Lewis from the book *Soldiers of Florida*. https://myweb.fsu.edu/rthompson2/cw/2-fl-cav/2-fl-cav-k.html. Accessed 24 June 2022.

Florida Department of State. "County Name Origins." Division of Historical Resources. www.flheritage.com/facts/reports/names/county.cfm. Accessed 24 June 2022.

———. "Florida in the Civil War." Florida Museum of History. https://www.museumoffloridahistory.com/exhibits/permanent-exhibits/florida-in-the-civil-war/. Accessed 24 June 2022.

"General William Birney, USA." https://www.historycentral.com/Bio/Obama/Bio/people/UGENS/USABirney.html. Accessed 24 June 2022.

Grzelak, Jeff. "Five Fateful Days in February 1865: The 'Cotton Raid' and Ambush at Braddock's Farm." 17thcvi.org. An Online History of the 17th Connecticut Volunteer Infantry during the U.S. Civil War. http://seventeenthcvi.org/blog/engagements/braddocks-farm. Accessed 24 June 2022.

Harrison, Donald H. "The Two Presidents Johnson and the Jews." San Diego Jewish World. 31 October 2012. https://www.sdjewishworld.com/2012/10/31/the-two-presidents-johnson-and-the-jews/.

Hickox, O. J., Jr. "A Brief History of Clinch's Regiment, 4th Georgia Volunteer Cavalry." March 2011. Coastal Georgia Genealogy & History. www.glynngen.com/military/civilwar/4thga/DLClinch.htm.

Zerfas, Lewis L. "The Hillsborough River Raid and the Battle of Ballast Point." *America's Civil War* (November 2001).

"Historic Markers Across Florida." http://apps.flheritage.com/markers/.

"Historic Peace River Valley, Florida: Battle of Bowlegs Creek." 30 April 2011. http://peacerivervalley.blogspot.com/2011/04/battle-of-bowlegs-creek-fort-meade.html.

"History of Gainesville, Florida." George A. Smathers Libraries Digital Collection. https://ufdc.ufl.edu/UF00016410/00001. Accessed 24 June 2022.

Hollway, Don. "Chasing Jefferson Davis." Warfare History Network. Winter 2017. https://warfarehistorynetwork.com/daily/civil-war-chasing-jefferson-davis.

House, Art. "Warren History Background." An Online History of the 17th Connecticut Volunteer Infantry during the U.S. Civil War. https://seventeenthcvi.org/blog/history-index/william-warren-history/. Accessed 27 June 2022.

"Jacob Summerlin: The Cowman who was King of Crackers." 21 August 2007. https://www.tbnweekly.com/pinellas_county/article_7c5bbe73-4029-57b6-a427-0ab5758c30ca.html.

"James Aldridge Braddock." Founding Florida Pioneers Settlers & Their Descendants. 14 May 2013. https://www.genealogy.com/ftm/m/i/z/Jean-H-Mizell/PHOTO/0012photo.html

"Johnston Surrenders at Bennett Place." https://www.battlefields.org/learn/articles/bennett-place-surrender.

Kozikowski, Kara A. "Guerrilla Warfare: Hometown Heroes and Villains." American Battlefield Trust. https://www.battlefields.org/learn/articles/guerrilla-warfare. Accessed 24 June 2022.

Thiesen, William H. "The Long Blue Line: Lighthouse Service during the Civil War."

Martin, James. "Civil War Conscription Laws." *In Custodia Legis*: Law Librarians of Congress. 15 November 2012. https://blogs.loc.gov/law/2012/11/civil-war-conscription-laws.

Martin, Richard A. "The Great River War on the St. Johns." http://mapleleafshipwreck.com/Book/Chapter2/chapter2.htm. Accessed 24 June 2022.

Milton Letterbook. State Archives of Florida, Florida Memory. 23 April 1862 https://www.floridamemory.com/items/show/266377; 25 April 1862; 13 March 1862; 3 April 1862; 24 March 1862. Accessed 27 June 2022.

Murphree, R. Boyd. "Florida and the Civil War: A Short History." A Guide to Civil War Records at the State Archives of Florida. See paragraph heading "Florida Unionist." https://www.floridamemory.com/learn/research-tools/guides/civilwarguide/history.php. Accessed 24 June 2022.

"Our History." The Islands of the Bahamas. https://www.bahamas.com/our-history. Accessed 27 June 2022.

"Overview...Waccasassa, Wekiva, and Otter Creek Paddling." Paddle Florida. https://www.paddleflorida.net/waccasassa-wekiva-paddle.htm. Accessed 27 June 2022.

Phillips, Christopher. "Montgomery, James." Civil War on the Western Border: The Missouri-Kansas Conflict, 1855–1865. Kansas City [MO] Public Library Digital History. https://civilwaronthewesternborder.org/encyclopedia/montgomery-james. Accessed 24 June 2022.

"Preserving Florida's Railroad History." https://www.frrm.org/history/.

Riley, Darrell G. "The Civil War Years, 1861–1865." *Ocala Star Banner/Ocala.com*. Updated 22 September 2003. https://www.ocala.com/story/news/2003/01/01/the-civil-war-years/31271366007/.

"St. Augustine, Florida: Demographics." https://en.wikipedia.org/wiki/St._Augustine,_Florida#Demographics. Accessed 27 June 2022.

Starkey, Ty. "1st Florida Special Cavalry." https://rebelyell71.tripod.com/1st_fl_spec_cav.htm. Accessed 27 June 2022.

Stone, Spessard. "The Civil War in the Lower Peace River Valley." 25 Nov 1999. https://freepages.rootsweb.com/~crackerbarrel/genealogy/Civilwar3.html.

"Thomas Wentworth Higginson." American Battlefield Trust. https://www.battlefields.org/learn/biographies/thomas-wentworth-higginson. Accessed 27 June 2022.

UHP Staff. "Caught in the Middle: New Smyrna." Ultimate History Project. www.ultimatehistoryproject.com/caught-in-the-middle.html. Accessed 24 June 2022.

University of South Florida [System]. University Communications and Marketing. "100 Years of Black St. Petersburg" (1998). USF Magazine Articles about USF St. Petersburg campus. 27. https://scholarcommons.usf.edu/usf_mag_articles_usfsp/27. Cited as USF, "100 Years of Black St. Petersburg."

Voyles, Karen. "Remembering former Fla. governor Perry." *Gainesville Sun*. 13 November 2011. https://www.gainesville.com/story/news/2011/11/13/remembering-former-fla-governor-perry/31822361007/.

Williamson, Ronald. "Birney's Raid in the News." 28 April 2007. https://www.treasurenet.com/threads/birney-s-raid-in-the-news.45480/

Yeats, William Butler. "The Second Coming." https://www.poetryfoundation.org/poems/43290/the-second-coming. Accessed 27 June 2022.

Index